T0234734

Practical RF Amplifier Design and Performance Optimization with SPICE and Load- and Source-pull Techniques

Amal Banerjee

Practical RF Amplifier Design and Performance Optimization with SPICE and Load- and Source-pull Techniques

 Springer

Amal Banerjee
Analog Electronics
Kolkata, West Bengal, India

ISBN 978-3-030-62514-6   ISBN 978-3-030-62512-2   (eBook)
https://doi.org/10.1007/978-3-030-62512-2

This Springer imprint is published by the registered company Springer Nature Switzerland AG
The registered company address is: Gewerbestrasse 11, 6330 Cham, Switzerland

This book is dedicated to:
My late father Sivadas Banerjee
My mother Meera Banerjee
My sister Anuradha Datta
A dear friend, mentor and guide
Dr. Andreas Gerstlauer
The two professors who taught me the basics
of radio frequency physics and electronics
Dr. Melvin E. Oakes
Dr. C. Fred Moore

Contents

Chapter 1
Introduction and Problem Statement

1.1 Topic Definition

A radio frequency (RF) amplifier is an electronic circuit that is used in both wired and wireless telecommunication devices, networks, and systems to compensate for unavoidable signal energy loss between the signal transmitter and signal receiver. This book explains, with an exhaustive set of design examples, how to design and then evaluate and optimize the performance characteristics of such an RF power amplifier, without the help of any expensive, high-end test equipment and/or expensive CAD (Computer Aided Design) software tools. This is to guarantee that the amplifier works exactly as per specifications. Given that telecommunication networks and systems perform mission critical operations in each corner of our planet, it is absolutely essential that such telecommunication systems work exactly as per predefined specifications; or in short, the components of a telecommunication network or system must work exactly right within acceptable predefined tolerances.

This book explains and demonstrates with an exhaustive set of design examples, how common types of radio frequency (RF) amplifiers (classes A, B, AB, C, D, E, F, G, and H) can be designed, and then have their performance characteristics evaluated and optimized. The key performance evaluation task is performed with the most accurate, reliable electrical/electronic circuit simulation and performance evaluation tool SPICE [15] (Simulation Program with Integrated Circuit Emphasis) used along with the electronics industry standard load, source pull techniques *WITHOUT* using any expensive test equipment (spectrum, network analyzers, etc.,) or high-end CAD (Computer Aided Design) tools (Keysight Technologies ADS [19, 20], Microwave Office AWR [21], Cadence Virtuoso SpectreRF [22] etc.).

Radio frequency (RF) amplifiers have been designed, constructed, and tested since early days of radio transmission, and RF amplifier design has been a favorite topic of electronics engineering text and specialized book authors [1–16, 28]. However, there are serious drawbacks.

© The Author(s), under exclusive license to Springer Nature Switzerland AG 2021
A. Banerjee, *Practical RF Amplifier Design and Performance Optimization
with SPICE and Load- and Source-pull Techniques*,
https://doi.org/10.1007/978-3-030-62512-2_1

- The explanations and treatment is too high level or mathematical or a combination of both or focused on a few amplifier classes and types. The reader, unless intimately familiar with the topic, finds it difficult to extract usable design equations or understand the sequence of design calculation steps required to complete a design. This leads to incorrect designs [10], in which case a shunt capacitor (part of the output impedance matching network) short circuits the output signal to ground.
- The rudimentary design examples use ideal switches [1–16, 28] as semiconductor devices (bi-junction or field-effect transistor). This is incorrect as all semiconductor devices (starting with the basic p-n junction diode) show clear *nonlinear behavior*. All SPICE [18, 23–26] versions include nonlinear semiconductor device models, freely available from reputed semiconductor device manufacturers. These SPICE [18, 23–26] semiconductor device models can be used with any available open-source or proprietary SPICE [18, 23–26] simulator.
- The literature [1–16, 28] contains a few bare-bones design examples that state the main results, but no details about the sequence of calculation steps used to obtain those results. The reader is left confused as ever.
- In keeping with the ideal switch model of semiconductor devices, some book (text/specialized) authors use the small signal S parameters, supplied (by the semiconductor device manufacturer) with the selected transistor's data sheet. However, steady state behavior can only be modeled with large signal S parameters, and so these rudimentary amplifier design examples are incomplete and inaccurate. *SPICE's* [18, 23–26] *transient analysis feature fully exploits the underlying nonlinear behavior of a transistor as well as being inherently large signal.*
- *None of the electronic amplifier text/specialized books* [1–16, 28] *provide any details (methods, procedures) about performance evaluation of a given RF power amplifier, let alone use the electronics industry standard load and source pull techniques that accurately estimate source and load impedances at the input and output ports of the amplifier for correct impedance matching.*
- Although load and source pull schemes have been incorporated into available CAD tools [19–22], these use SPICE [18, 23–26] as the main circuit simulation engine for performance evaluation of the test electronic circuit. Unfortunately, all these CAD tools are very expensive and have steep learning curves.

1.2   Solution

The novel scheme to address the issues elaborated on earlier exploits the full power of the most accurate, reliable, rugged, and thus universally accepted electrical/ electronic circuit simulation and performance evaluation tool SPICE [18, 23–26]. *Specifically, the new scheme exploits the transient analysis feature of SPICE* [18, 23–26] *to accurately simulate and extract the performance characteristic of the steady state behavior of RF power amplifiers. It also exploits the freely available and extensively used SPICE* [18, 23–26] *transistor models. The values of the passive components (capacitor, inductor, and resistor) for each class of RF power amplifier (e.g., class B) are computed with a supplied ANSI C computer language* [27] *executable. A C computer language* [27] *executable is supplied for each class (A, B, C, AB, D, E, F, G, and H), and versions for both the popular Linus and Windows operating systems are provided. The advantages of this approach are huge.*

- A supplied set of C computer language [27] executable programs (one set each for the popular Linux and Windows operating systems) compute the passive component (capacitor, inductor, and resistor) values for each RF power amplifier type (e.g., class C). **That is, class A has a dedicated C computer language** [27] **executable to compute the passive component values for any class A amplifier and so on for each amplifier class.** *The results are arranged in a text SPICE* [18, 23–26] *input format netlist. This eliminates all calculation errors (that might creep in if the values were computed manually), and the text SPICE* [18, 23–26] *input format netlist can be easily edited for design space exploration.*
- Any available SPICE [18, 23–26] open-source or proprietary simulator can be used, and the transient analysis results are in text comma, tab separated variable format that can be easily processed by other programs to compute, e.g., RMS (root mean square) amplifier load current.
- The electronics industry standard load and source pull schemes can be easily included in the text SPICE [18, 23–26] netlist enabling design space exploration.
- The SPICE [18, 23–26] semiconductor device (bi-junction or field effect transistor) model can be easily and freely downloaded from the Web sites of reputed semiconductor device manufacturers such as Infineon Technologies, On Semiconductor, Vishay, and NXP Semiconductor. **This eliminates inaccuracies in simulation results.**
- **The SPICE** [18, 23–26] **simulation engine uses its built-in powerful** *transient analysis* **method to measure the steady state performance characteristics of the RF power amplifier under test. It also uses its built-in noise analysis feature to estimate the input and output noise spectra of the RF power amplifier being analyzed.**
- There is no need to use any expensive and difficult-to-master CAD (Computer Aided Design) tools such as AWR [19–22].

In the next chapter, existing commonly used radio frequency (RF) power amplifiers are examined in detail, specifically their classification (linear, nonlinear), design calculation steps and procedures, performance characteristics, and the electronics industry standard load and source pull [17] schemes.

In the final key chapter, an exhaustive set of RF power amplifier design examples are presented and analyzed in minute detail. While a set of supplied C computer language [27] *executables (one each for each RF power amplifier class, eight in total, a set each for both the popular Linux and Windows operating systems) are used to compute the passive component (capacitor, inductor, and resistor) values for each of these RF power amplifier classes, the analysis and performance characteristics (efficiency, input, output noise spectrum) of each RF power amplifier design are done with Ngspice 31 (any available open-source or proprietary SPICE [18, 23–26] distribution can be used). The Ngspice 31 simulator uses its transient analysis feature to analyze the steady state performance characteristics of each RF amplifier and the built-in noise analysis feature to measure the test RF amplifier's input and output noise spectrum.*

References

1. Grebennikov A (2015) RF and microwave power amplifier design, 2nd edn. Mc-Graw Hill Educational. ISBN 978-07-0-182862-8
2. Walker J (ed) (2012) Handbook of RF and microwave power amplifier. Cambridge University Press. ISBN 978-0-521-76010-2
3. Bahl I. Fundamentals of RF and microwave transistor amplifiers. John Wiley and Sons. ISBN 978-0-470-39166-2
4. Grebennikov A, Kumar N, Binboga SY (2017) Broadband RF and microwave amplifiers. CRC Press. ISBN 9781138800205 - CAT# K32788
5. Eroglu A (2015) Introduction to RF power amplifier design and simulation. CRC Press. ISBN 978-1-4822-3165-6
6. Cripps SC (2006) RF power amplifiers for wireless communications. Artech House. ISBN 10: 1-59693-018-7. ISBN 13: 978-1-59693-018-7
7. Cripps SC (2002) Advanced techniques in RF power amplifier design. Artech House Print on Demand. ISBN 10: 1580532829. ISBN 13: 9781580532822
8. Sokal's original class E amplifier paper from: https://people.physics.anu.edu.au/~dxt103/160m/class_E_amplifier_design.pdf
9. Another source for Sokal's paper: https://people.eecs.berkeley.edu/~culler/AIIT/papers/radio/Sokal%20AACD5-poweramps.pdf
10. Slade G. Notes on designing class E RF amplifiers. https://www.researchgate.net/publication/320623200_Notes_on_designing_Class_E_RF_power_amplifiAers
11. Albulet M (2001) RF power amplifiers. Noble Publishing. ISBN 1-884932-12-6
12. Kazimierczuk MK (2015) RF power amplifiers, 2nd edn. John Wiley & Sons. Print ISBN 9781118844304. Online ISBN 9781118844373A
13. Shirvani A, Wooley BA. Design and control of RF power amplifiers. Springer. Ebook ISBN 978-1-4757-3754-7. Hardcover ISBN 978-1-4020-7562-9
14. Rudiakova AN, Krizhanovski V (2006) Advanced design techniques for RF power amplifiers. Springer. ISBN 978-1-4020-4639-1
15. Pozar DM. Microwave engineering, 4th edn. John Wiley and Sons Publication. ISBN 978-0-470-63155-3
16. Colantonio P, Giannini F, Limiti E. High efficiency RF and microwave solid state power amplifiers. John Wiley and Sons. Print ISBN 9780470513002. Online ISBN 9780470746547. https://doi.org/10.1002/9780470746547
17. Good introduction on load pull from: http://mwrf.com/test-measurement/impedance-tuning-101
18. Ngspice users' manual for the latest version (31) may be downloaded easily and freely. http://ngspice.sourceforge.net/docs/ngspice-manual.pdf
19. Keysight technologies advanced design system users manual from: https://edadocs.software.keysight.com/display/ads201101/PDF+Files+for+ADS+Documentation
20. How to design an RF power amplifier class E. https://www.youtube.com/watch?v=iABwHeZ3_Jw

21. AWR Microwave Office. https://www.awr.com/software/products/microwave-office
22. Cadence Virtuoso SpectreRF. https://www.cadence.com/zh_TW/home/training/all-courses/84474.htm
23. Latest Ngspice version 31 user guide and manual from: http://ngspice.sourceforge.net/docs/ngspice-manual.pdf
24. LTSpice users guide and manual from: https://ecee.colorado.edu/~mathys/ecen1400/pdf/scad3.pdf
25. Pspice users guide and manual from: https://www.seas.upenn.edu/~jan/spice/PSpice_UserguideOrCAD.pdf
26. HSpice users guide and manual from: https://cseweb.ucsd.edu/classes/wi10/cse241a/assign/hspice_sa.pdf
27. Ebook of the all-time classic C programming language book by the creators of the C computer language Brian Kernighan and Dennis Ritchie, can be downloaded easily from: http://www2.cs.uregina.ca/~hilder/cs833/Other%20Reference%20Materials/The%20C%20Programming%20Language.pdf
28. Grebennikov A, Kumar N (2015) Distributed RF power amplifiers for RF and microwave communications. Artech House. ISBN 9781608078318

Chapter 2
Radio Frequency Power Amplifier (Narrow Band, Distributed) Design Fundamentals: Design Procedures and Analysis

2.1 Introduction

Ideally, a radio frequency (RF) power amplifier, a two-port electronic circuit converts steady state DC power supply electrical power into oscillatory or sinusoidal electrical power at the output. Effectively, a radio frequency amplifier converts low-power input signal to a high-power output signal. The input signal applied at the base (bi-junction transistor) or gate (field effect transistor) acts as a control or trigger. An RF amplifier is a key component of any modern signal processing circuit or system, e.g., wireless transmitter or receiver, since unavoidable physical effects degrade electronic signal power, and the lost signal energy must be compensated to guarantee error-free signal reception. So, RF power amplifiers must operate within tight tolerances on the design specifications and the circuit designer must be able to estimate performance characteristics before the physical device is fabricated. This is a challenging task.

This book examines in detail (with an exhaustive set of design examples) a novel scheme to design and estimate the performance characteristics of some common narrow band and distributed RF power amplifiers. *It exploits the simple, yet extremely rugged and powerful C [22] computer language and the most reliable, trusted, and widely used electrical/electronic circuit simulation tool SPICE (Simulation Program with Integrated Circuit Emphasis) [18–21]. The C computer language executables accept appropriate user-specified input parameters (e.g., bi-junction transistor collector-emitter voltage), compute the amplifier circuit passive (capacitor, inductor, resistor) component values, and finally generate the text SPICE [18–21] input format netlist, which can be used with any available open-source or proprietary SPICE [18–21] simulator.* The SPICE [18–21] simulator uses its *transient analysis* (large signal) feature to generate a text file output. The contents of this output text file are processed by another utility C computer language [22] executable *rmscalc* that calculates the RMS (root mean square) values from the input

A. Banerjee, *Practical RF Amplifier Design and Performance Optimization with SPICE and Load- and Source-pull Techniques*, https://doi.org/10.1007/978-3-030-62512-2_2

data which are then used to calculate the selected RF amplifier performance characteristic, e.g., efficiency. SPICE [18–21] is the most accurate, reliable electrical/electronic circuit simulator, guaranteeing accuracy and reliability of its in-built analysis, e.g., transient analysis. Most importantly, SPICE [18–21] uses bi-junction and field effect transistor device models that are obtained by nonlinear curve fit of test data provided by semiconductor device manufacturers. So, inaccuracies in the SPICE [18–21] transient analysis result as a consequence of using ideal switch models as transistors are eliminated. These device models are freely and readily available from semiconductor device manufacturer's Web sites. The text SPICE [18–21] input format netlist can be edited easily, which allows design space exploration and design optimization.

2.2   RF Power Amplifier Types (Linear, Nonlinear), Classes (A, B, AB, C, D, E, F, G, and H), and Conduction Angle

A linear RF power amplifier generates *output signal that is proportional to the input signal*, but the output signal power is much higher than the input signal power – hence the name *RF power amplifier*. A key concept directly related to the linear RF power amplifier is *conduction angle* and is derived from analyzing the output with a pure sine wave input. If the transistor is always on, the conducting angle is 360°. If it is on for only half of each cycle, the conduction angle is 180°. For a nonlinear RF power amplifier, *a change in the input signal **does not produce proportional change** in the output signal*. Moreover, for a nonlinear RF power amplifier, the active device (bi-junction or field effect transistor) acts only as a switch with a turn-on resistance R_{ON}, and *no transistor biasing is required*.

Traditionally, RF power amplifiers have been named as A, B, C, AB, D, E, F, G, and H, and so on. Of these, A, B, C, and AB are linear and D, E, and F are nonlinear. Classes G and H use the linear class AB RF power amplifier to perform the amplification task. The linear amplifier classes are classified by the length of their conduction state over some portion of the output waveform, such that the output stage transistor operation lies somewhere between being "fully-ON" and "fully-OFF." Non-linear amplifiers operate such that the transistor is saturated ("fully-ON") ot cut-off ("fully-OFF"), no intermediate state.

Both linear and nonlinear amplifiers are essentially narrow band in nature, and special circuit features are required to ensure that their gain (ratio of output signal power to input signal power) remains constant (within applicable tolerances). *Distributed amplifiers* address this issue, i.e., maintain a constant gain (within applicable tolerances) over a very wide frequency band (often reaching hundreds of GHz).

2.3   Key Load Line Concept and Biased (Linear) RF Power Amplifier Design

The load line concept is specific to the design and analysis of any nonlinear electronic circuit (one that contains at least one nonlinear semiconductor device). *It is a line drawn on the characteristic curve (curve of current versus voltage – e.g., for bi-junction transistor I_{CE}, V_{CE} curve) and* **denotes the response of the linear part of the circuit to the semiconductor device.** *The intersection of any load line with any semiconductor device characteristic curve is called the Q point and indicate that current (through) and voltage (across)* **both the semiconductor device and the external circuits match.** *For a bi-junction transistor, any I_{CE}, V_{CE} curve depends on the base emitter current I_{BE} and so having a load line on a series of I_{CE}, V_{CE} curves (each depending on a specific I_{BE}) shows how I_{BE} will affect the operating or Q point.*

Load lines can be used separately for both DC and AC analyses. Bi-junction transistors used in linear RF power amplifiers have both DC and AC currents flowing through them. A DC current source biases it to the correct operating point(Q) and an AC signal applied at the transistor base is superimposed on the DC biasing voltage. The DC load line is the load line of the DC-equivalent circuit. The DC-equivalent circuit is defined by replacing capacitors by open circuits and inductors by short circuits, i.e., capacitive reactance is infinite and inductive reactance is zero. This determines the correct DC operating point – the Q point.

Now, an AC load line can be drawn through the same Q point. The AC load line is a straight line with a slope equal to the AC impedance facing the transistor. The AC load line represents the ratio of AC voltage to current in the transistor. As both capacitive and inductive reactances are frequency dependent, the slope of the AC load line varies with the frequency of the applied signal, thereby creating a family of AC load lines, each dependent on a separate operating frequency. The limiting AC load line corresponds to operating the circuit at "infinite frequency" and is equivalent to short-circuited capacitors and open inductors.

The load line is used to design all three linear RF amplifiers A, B, and C, as shown in Fig. 2.1. The remainder of this chapter is devoted to the discussion of these common linear and nonlinear RF power amplifiers. The section on each amplifier

Fig. 2.1 Load line for generic RF NPN transistor, with 'Q' points for classes A, B, and C

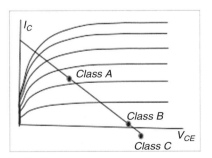

type, e.g., class A, contains a brief description of that amplifier's circuit, followed by relevant design equations and formulas; the details of derivation of these are in [1–16]. All subsequent discussions are based on the generic NPN RF transistor.

2.3.1 Biased (Linear) Class A (Common Emitter) RF Power Amplifier Load Line, Design Equations, and Analysis

The class A amplifier is the familiar common emitter amplifier, as shown in Fig. 2.2. The operating "Q" point is selected to lie in the middle of the I_{CE}, V_{CE} graph as shown in Fig. 2.1. The design equations are the same as that for the common emitter amplifier. The values for the DC blocking capacitors are chosen to minimize AC resistance. The designer supplies the following input parameters (available from transistor data sheet).

- Minimum transistor forward DC current gain h_{FE} or *beta*
- Collector current I_C for minimum $h_{FE, MIN}$
- Target load output power P_{OUT}
- Operating frequency f_o
- The DC power supply voltage V_{CC}

The DC blocking capacitor is $\frac{1}{f_o}$

The collector resistor is $R_C = \frac{0.5\,V_{CC}}{I_C}$ and the emitter resistance is $R_E = \frac{0.1\,V_{CC}}{I_C}$. The emitter bypass capacitor's reactance at the operating frequency is $X_{EMIT,CAP,BYP} = 10.0\,R_E = \frac{V_{CC}}{I_C}$, so that the bypass capacitor's value is $C_{EMITTER,BYP} = \frac{I_C}{6.28\,f_o\,V_{CC}}$. The base emitter voltage is $V_{BE} = \frac{0.1\,V_{CC}}{I_C} + 0.7$ and the first base bias resistor value is $R_{BI} = \frac{V_{CC}}{\frac{10.0\,I_C}{h_{FE,MIN}}}$. The second base bias resistor value is obtained from the familiar voltage divider formula and the known value for R_{B1}, i.e., $\frac{R_{B2}}{R_{BI}+R_{B2}} = \frac{V_{BE}}{V_{CC}}$, so that $R_{B2} = \frac{V_{BE}R_{BI}}{V_{CC}-V_{BE}}$. The radio frequency choke at the transistor collector has a reactance $6.28\,f_o L = 10.0\,I_C R_C$ at the operating frequency f_o, so that the radio frequency choke inductance has a value $L_{RFC} = \frac{10.0\,I_C R_C}{6.28\,f_o}$. The initial estimate of the load

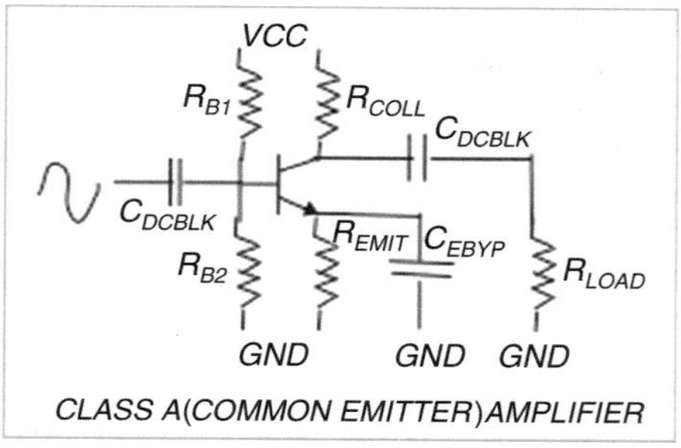

Fig. 2.2 Generic class A RF power amplifier

CLASS A(COMMON EMITTER)AMPLIFIER

impedance is $R_L = \frac{V_{CC}^2}{2.0\,P_{OUT}}$. This value of the load impedance is an initial value, as it can be varied using the load and/or source pull technique to optimize amplifier performance characteristics. This will be elaborated in subsequent sections. *The conduction angle is 360°, as the bias network ensures that the transistor is always switched on, even in the absence of signal, reducing efficiency, whose theoretical maximum limit is 50%.* A design example for this amplifier is presented in the next chapter.

2.3.2   Biased (Linear) Single-, Double-Sided Class B RF Power Amplifier Load Line, Design Equations, and Analysis

Although the class B amplifier is a biased (linear) power amplifier, the transistor base for both the single- and double-sided (Fig. 2.3a, b) versions require negligible biasing, in practice bias voltage is set to zero (*self-biasing*). *So, the transistor switches on only when the input signal voltage exceeds the base – emitter diode voltage drop of 0.7 V. This improves the efficiency (compared to the class A amplifier) as the transistor draws less power. The maximum theoretical efficiency of the class B amplifier is approximately 80%.* The single-sided class B amplifier is used only for RF applications; the double-sided class B amplifier, *with the resonator*, is used for RF applications as well. The input DC blocking capacitor is eliminated by using an input coupling transformer and for the single sided class B amplifier, the output DC blocking capacitor is used only if the resonator consists of a parallel capacitor, inductor pair, for a series resonator, the resonator capacitor also serves a DC blocking capacitor. *If a parallel resonator is used, both the capacitor and inductor values must be selected to provide maximum reactance, and if a series resonator is used, both capacitor and inductor must be chosen to provide minimum reactance.* Both *these guarantee that maximum output signal energy is delivered to the load.* The resonator is very important for the single-sided class B amplifier, as the transistor switches on only during the positive polarity portion of the

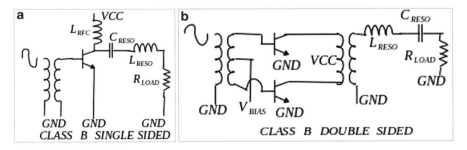

Fig. 2.3 (**a**) Single-sided class B amplifier. (**b**) Double-sided class B amplifier

input signal, and the resonator compensates for the negative or zero polarity portion of the input signal. The double-sided class B amplifier, if used *without the resonator*, suffers from *cross-over distortion* problem, i.e., the transistor switches off when the input signal level falls below the diode drop. The input and output transformer inductors are selected to provide sufficient AC resistance so that only a negligible part of the signal energy leaks to ground – this requires some experimentation and design space exploration.

At the operating frequency f_o, to satisfy the resonance condition, $f_o = \frac{1.0}{6.28\sqrt{C_{RESO}\,L_{RESO}}}$ or $C_{RESO}\,L_{RESO} = \frac{1.0}{39.44 f_o^2}$. So, at a predefined operating frequency, f_o, the values for the resonator (parallel, series) can be easily estimated, such that they satisfy either the maximum or minimum AC resistance condition as mentioned earlier. The initial value of the load impedance is computed as $R_L = \frac{V_{CC}^2}{2.0\,P_{OUT}}$. The single-sided and double-sided class B RF amplifiers are shown in Fig. 2.3a, b. At design time, the operating frequency, target load output power, and the transistor's collector-emitter voltage (from manufacturer supplied data sheet) are required. Design examples are provided in the next chapter.

2.3.3 Biased (Linear) Single-Sided Class C RF Power Amplifier (Fixed, Zero Bias) Load Line, Design Equations, and Analysis

The class C RF power amplifier is biased (linear) amplifier, which comes in two varieties, that is, *fixed* and *zero (self)* bias. For the fixed-bias case, a negative low-value bias is applied to the transistor base. The class C amplifier has been conceived of to maximize efficiency. This means that with a fixed-bias version, the input signal mist exceed both the base – emitter diode drop and the negative external bias for the transistor to switch on. *This ensures that the transistor is switched on only when the input signal level is higher than the bias level, thereby drawing less power and forcing the conduction angle to be less than 180°. Identical arguments hold for the zero-bias case, although the bias level is just the base – emitter diode drop. Both fixed and zero-bias versions boost efficiency above both class A and B amplifiers, theoretically reaching 100%.* For both the fixed and zero-bias cases, an output DC blocking capacitor is required, as the parallel resonator is connected directly to the transistor collector. The input DC blocking capacitor may be replaced with a coupling transformer, only in the zero base bias version. At an operating frequency f_o, the DC blocking capacitor is $\frac{1.0}{f_o}$, and the resonator C, L product is $C_{RESO}L_{RESO} = \frac{1.0}{39.44 f_o^2}$. The two versions of the amplifiers are shown in Fig. 2.4a, b. As in the case of the class B amplifier, the designer has to supply the operating frequency, the collector-emitter voltage, and the target load output power.

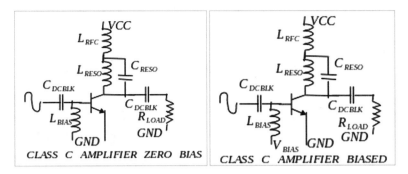

Fig. 2.4 (**a**) Zero-base bias class C amplifier. (**b**) Fixed-base bias class C amplifier

2.3.4 Biased (Linear) Class AB RF Power Amplifier Design Equations and Analysis

The class AB amplifier has been conceived of to take advantage of the best features of the class A and B amplifiers and also eliminate their deficiencies. It does not require any resonator. It can be used for RF applications if the candidate NPN transistor has a PNP counterpart such that their transition frequencies ($f_{T, NPN}f_{T, PNP}$) lie in the RF frequency range (hundreds of MHz to tens of GHz). It overcomes the *cross-over distortion* issue of the double-sided class B amplifier by having the PNP transistor, and transistor biasing is achieved by a biasing network consisting of resistors and diodes. A class AB amplifier is biased so that output current flows for less than one full cycle of the input waveform but more than a half cycle. The circuit is similar to the double-sided class B amplifier – two transistors form the complementary output stage so that each transistor conducts on opposite half-cycles of the input signal. **So, by forcing both transistors to conduct current at the same time for a very brief period of time, during the input signal zero crossing, the cross-over distortion issue of the class B amplifier is curtailed.** The conduction angle is greater than 180° but less than 360°. The class AB amplifier is more efficient than class A amplifier but less efficient than class B – *a small quiescent current needed to bias the transistors just above cut-off.* The calculation of the base bias and collector resistors (for both NPN and PNP transistors) uses the same equations as those for the class A amplifier. The collector resistor, for the NPN transistor, for a given selected value of $h_{MIN, NPN}$, and corresponding collector current, is $\frac{V_{CC,NPN}}{I_{C,NPN}}$ and similar expression holds for the PNP transistor. The base bias resistor for the NPN transistor is $\frac{V_{CC,NPN}}{10.0\ I_{C,NPN}}\big/{h_{FE,MIN,NPN}}$. Similar expression holds for the PNP transistor. Each of the input/output DC blocking capacitors operating at frequency f_o has numerical value $\frac{1}{f_o}F$, where F denotes Farads, the unit of capacitance. The class AB amplifier circuit is shown in Fig. 2.5.

Fig. 2.5 Class AB amplifier

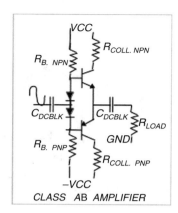

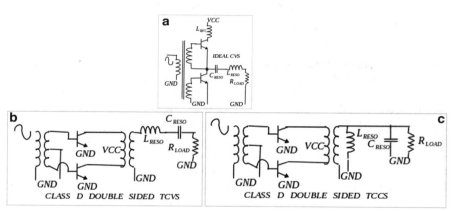

Fig. 2.6 (**a**) Idealized complementary voltage switching. (**b**) Transformer-coupled voltage source class D RF power amplifier. (**c**) Transformer-coupled current source class D RF power amplifier

2.3.5   *Switched (Nonlinear) Class D Double-Sided RF Power Amplifier Design Equations and Analysis*

The class D RF power amplifier is a switched amplifier, in which the transistor is used as a switch with a turn-on resistance R_{ON}. *The transistor is either in the fully saturated state or in a cut-off state, and the output signal is not proportional to the input.* It exists in several versions, starting from transformer-coupled current switch (TCCS), transformer-coupled voltage switch (TCVS), and so on – only the first two that are commonly used will be examined in detail here.

To understand the operation of this amplifier, consider the operation of the idealized complementary voltage source class D amplifier (CVS – Fig. 2.6a). The input transformer provides the base trigger signal to the two transistor bases, depending on the polarity of the input signal. As each transistor is either in the saturated (ON state) or in cut-off (OFF state) state, the pair forms a two-pole switch that connects the series resonator alternatively to the DC power supply or to the ground.

This idealized circuit is based on the assumption that both transistors and the capacitor and inductors are ideal devices with no parasitic effects.

The transformer-coupled voltage source class D amplifier (Fig. 2.6b) is a practical extension of CVS. In case the lower transistor is switched on while the upper transistor is switched off, the voltage across the lower half (bifilar winding of output transformer) is V_{CC} and the voltage across the upper half of the same transformer winding is $-V_{CC}$. When the upper transistor is switched on, and the lower transistor switched off, both halves of the output transformer have the same (magnitude) voltage across them, but of opposite polarity. Then the other winding gets a voltage $\pm V_{CC}$ assuming a 1:1 transformer winding ratio.

For the transformer-coupled current source (TCCS) class D RF power amplifier (Fig. 2.6c), the series resonator is replaced by a parallel resonator with one terminal of both the resonator capacitor and inductor grounded. The output at the output transformer secondary is an alternating polarity current signal.

The above analysis is based on the assumption that all devices are ideal. The best way to design, analyze, and estimate performance characteristics of these amplifiers is to use the best electrical/electronic circuit simulator and performance evaluation tool SPICE (Simulation Program with Integrated Circuit Emphasis) [18–21]. The next chapter is dedicated to designing, analyzing, and estimating performance characteristics of each of these RF power amplifiers using SPICE [18–21]. Design equations are simple. At an operating frequency f_o, the product of the resonator capacitor and inductor is given by $C_{RESO} L_{RESO} = \frac{1.0}{39.44 f_o^2}$. If the TCVS version of the class D amplifier is used, values for the series resonator capacitor and inductor are selected to minimize both their reactances at f_o, and if a parallel shunted resonator is used, the reactances of both the capacitor and inductor are maximized. In both cases, the goal is to maximize output signal energy being dumped into the load.

2.3.6   Switched (Nonlinear) Class E RF Power Amplifier Design Equations and Analysis

The switched (nonlinear) class E [8, 9] amplifier was conceived of to boost the efficiency of the linear class B and C amplifiers. The transistor is used as a ON-OFF switch in a manner such that *simultaneous high current and high voltage in the transistor are prevented, thereby reducing power loss during switching transitions and boosting efficiency. To achieve this goal, the circuit must be arranged such that high current and voltage cannot exist at the same time.* Then the "target" current and voltage waves through the transistor during an RF cycle would appear as in Fig. 2.7a. This means that:

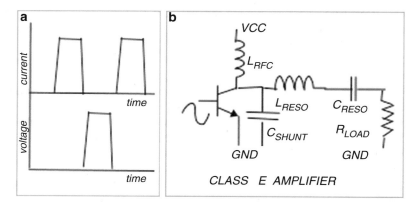

Fig. 2.7 (**a**) Current and voltage waves. (**b**) Class E RF power amplifier in ideal class E amplifier transistor

- Voltage rise through the transistor is prevented until the current through it is reduced to near zero and vice versa.

This goal is achieved with a suitable load network. Right at the moment of switch-on, the transistor *does not discharge* the shunt capacitor, that represents the transistor's output and any external capacitances (Fig. 2.7b).

Sokal [8, 9] originally used nonlinear curve fit to extract the values for the shunt capacitor, resonator capacitor, resonator inductor, and load impedance, using the following:

- Power supply voltage V_{CC}
- Collector radio frequency choke inductance value L_{RFC}
- Target output load power P_L
- Operating frequency f_o
- Circuit loaded quality factor Q_L
- Transistor saturation offset voltage V_o

The load impedance is

$$R_{LOAD} = \frac{0.576801 f(Q_L)(V_{CC} - V_o)^2}{P_L}, \text{where}$$

$$f(Q_L) = 1.000 - \frac{0.414395}{Q_L} - \frac{0.577501}{Q_L^2} + \frac{0.205967}{Q_L^3}$$

The shunt capacitor is

$$C_{SHUNT} = \left(\frac{1.0}{34.2219 f_o R}\right)\left(\frac{0.6}{(6.28 f_o)^2 L_{RFC}}\right) f(Q_L), \text{where}$$

$$f(Q_L) = 0.99866 + \frac{0.91424}{Q_L} - \frac{1.03175}{Q_L^2}$$

The resonator capacitor is

$$C_{RESO} = \left(\frac{1.0}{6.28 f_o R}\right)\left(\frac{1.0}{Q_L - 0.104823}\right)\left(1.00121 - \frac{1.01468}{Q_L - 1.7879}\right)$$
$$- \frac{0.2}{6.28 f_o^2 L_{RFC}}$$

Finally, the resonator inductor is

$$L_{RESO} = \frac{Q_L R}{6.28 f_o}$$

2.3.7 Switched (Nonlinear) Class F RF Power Amplifier Design Equations and Analysis

The switched (nonlinear) class F RF power amplifier is an improvement in the linear class B amplifier, by adding *harmonic tuning*. It is a well-known fact from Fourier analysis [23, 24] that a periodic time-varying current and voltage wave can always be expressed as a linear superposition of the odd harmonics of a sine current and voltage wave of fundamental frequency f_o. *Therefore, power amplifier efficiency can be boosted if, given an arbitrary (typically square wave) input trigger to an amplifier, the odd harmonics are filtered out, at the output port, i.e., amplifier acts as an open circuit for odd harmonics and a closed circuit for even harmonics.* The result is a class F amplifier. Mathematically, a square wave consists of a linear superposition of infinite number of odd harmonics – in reality, the contribution of high odd harmonics (7th, 9th, 11th, etc.) is small and hence ignored, and most RF class F power amplifiers are designed to tackle up to the seventh harmonic. A typical RF class F power amplifier with third and fifth harmonic tuning from the output waveform is shown in Fig. 2.8. *The parallel resonator for the fundamental f_o is an open circuit for the fundamental and closed circuit for **all other** harmonics, guaranteeing that maximum output signal energy at f_o is deposited on the load.* The input applied at the transistor base is a square wave at the fundamental frequency f_o, thereby guaranteeing that higher harmonics are generated at the output (transistor collector).

Fig. 2.8 Class F RF power
amplifier for third and fifth
harmonic tuning

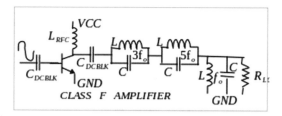

Designing the class F RF power amplifier involves computing the values for the capacitors and inductors from the odd harmonic resonators, up to the harmonic order to be tuned. Also, the capacitor and inductor value for the resonator for the fundamental must also be computed along with the input and output DC blocking capacitors. For the example class F amplifier of Fig. 2.8, for the third harmonic resonator, the product of the capacitor and inductor values is ($C_{RESO,3}L_{RESO,3} = \frac{1.0}{354.9456 \, f_o^2}$), allowing for the calculation of the values of these two components. Similarly, for the fifth harmonic resonator, $C_{RESO,5}L_{RESO,5} = \frac{1.0}{985.96 \, f_o^2}$. The DC blocking capacitors at the input and output must provide minimum capacitive reactance up to the fundamental f_o – the odd harmonics would be removed anyway, so it does not matter if the capacitive reactance for these is larger than that for the fundamental – $C_{DC,BLOCK} = \frac{1.0}{f_o}$.

2.3.8   Linear (Biased) Class G and H RF Power Amplifiers Design Equations and Analysis

The class G and H RF power amplifiers are class AB amplifiers with special circuitry to adjust both positive and negative power supply rails as per changes in input signal amplitude of both polarities. While the power conditioning circuitry for the class G amplifier switches between different power supply rails, class H modulates a fixed power supply rail (positive, negative) with the input signal, thereby providing variable power supply levels to the class AB amplifier. By definition, *Power = Volts × Amperes*, (where one or both the current and voltage values might be RMS (root mean square) values), the variable power supply rail feature immediately translates into efficiency boost. Effectively, both class G and H amplifiers are efficiency-boosted class AB amplifiers, i.e., *both are linear amplifiers*.

As the class AB amplifier has been examined and analyzed in detail in Sect. 2.3.4, the focus here is on variable rail power supplies. Audio equipment manufacturers have provided detailed online descriptions of the implementation of dual polarity variable rail power supplies, e.g., [27]. Detailed design examples of both class G and H amplifiers operating at RF frequencies have been presented in Chap. 3 of this book.

2.3.9   Biased (Linear) RF Power Amplifier Combination: Classes B and C Doherty Amplifiers

The Doherty amplifier [25, 26] was extensively used in radio transmitters in the early twentieth century, but it has recently seen a revival due to the worldwide expansion of wireless networks, especially cellular telephony. *The Doherty amplifier is used when high-efficiency RF power amplifiers are needed for amplifying signals with* **high peak to average power ratio**.

The Doherty amplifier power combines two amplifiers, the *carrier* amplifier and the *peaking* amplifier. The carrier amplifier is either a class AB (provides gain at any power level) or class B amplifier, while the peaking amplifier is class C which only operates at half of the cycle. This combination results in high power added efficiency (PAE). The input signal is split using a quadrature coupler, such as a Lange [15] or branchline hybrid [15].

The output of the Doherty amplifier involves careful signal processing. The two signals are out of phase by 90°, but by the addition of a quarter-wave transmission line of the peaking amplifier, they are brought back into phase and reactively combined. The two signals in parallel create a $\frac{Z_0}{2}$ impedance. This is stepped up to Z_0 by a quarter-wave impedance transformer [15]. In a 50.0 ohm system, the transformer would be 35.35 ohms. A typical Doherty amplifier configuration is shown in Fig. 2.9. Both the RF quadrature power splitter and the combiner are fabricated from transmission line microstrips and Z_0 is the characteristic impedance.

2.4   RF Power Amplifier Performance Metrics: Efficiency, Power-Added Efficiency, Gain, Noise Factor, Noise Figure, and Linearity

An RF power amplifier must satisfy initial design specifications, and so r*eliable and verifiable performance metrics and applicable tolerances must be defined to guarantee that a new design (that would translate to the actual physical circuit) satisfies the initial design specifications.* Specifying these performance metrics and applica-

Fig. 2.9 Doherty amplifier

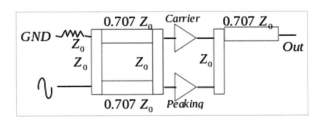

ble tolerances is essential, as both real-world active semiconductor devices (diodes, transistors) and passive (capacitors, inductors, and resistors) show non-ideal and nonlinear behavior.

The key RF power amplifier performance metric is efficiency, is defined as *the ratio of the supplied electrical energy from the power supply that is converted into useful output signal energy, per unit time*:

$$\text{Efficiency} = \frac{\text{Load power delivered (Watts)}}{\text{Electrical power supplied by power supply (Watts)}}$$

For each of the RF power amplifier classes examined so far, DC blocking capacitors or output coupling transformers have been used to guarantee that only AC electrical power is delivered to the load. So the above definition is refined to:

$$\text{Efficiency} = \frac{\text{RMS load power delivered (Watts)}}{\text{Electrical power supplied by power supply (Watts)}}$$

As will be seen in the SPICE [18–21] RF power amplifier design examples in the next chapter, electrical power delivered by the power supply is not purely DC in nature; therefore, the above definition for efficiency can be rewritten as:

$$\text{Efficiency} = \frac{\text{RMS load power delivered (Watts)}}{\text{RMS electrical power supplied by power supply (Watts)}}$$

The definition for power-added efficiency (PAE) is a simple modification in the definition for efficiency:

$$\text{PAE} = \frac{\text{RMS load power delivered (Watts)} - \text{RMS input signal power (Watts)}}{\text{RMS electrical power supplied by power supply } (Watts)}$$

The gain of a two-port network (e.g., RF power amplifier) measures by how much the signal output power is increased with respect to the input signal power. The units for gain for all RF electronic circuits are decibel (dB) and Neper (Np).

$\text{Gain(decibel)} = 10.0 \log\left(\frac{P_{output}}{P_{input}}\right)$ and $\text{gain}(N_p) = \frac{1}{2} \ln\left(\frac{P_{output}}{P_{input}}\right)$ where 'log' is logarithm to the base 10 and 'ln' is the natural logarithm to the base 2.71.

The *noise factor* of an RF power amplifier measures by how much the input signal is degraded by noise generated internally in the amplifier. It is defined as:

$F = \frac{\frac{S_{INPUT}}{N_{INPUT}}}{\frac{S_{OUTPUT}}{N_{OUTPUT}}}$ where S_{INPUT}, N_{INPUT} are respectively the signal and noise levels at the input and S_{OUTPUT}, N_{OUTPUT} are the signal and noise levels at the output. The *noise figure* is the noise factor F expressed in decibel (dB), $NF = 10.0 \log(F)$.

RF power amplifier linearity is how well it can amplify a signal without distortion, and this applies strictly to linear amplifiers (classes A, B, C, AB). This property is key for systems using modulation scheme *that encodes information in the amplitude variation of the signal*, ranging from simple Amplitude Modulation (AM) to complicated Quadrature Amplitude Modulation (QAM). Such a modulation scheme can only work if the receiver can accurately identify the differences in the incoming signal's amplitude, which in turn means a linear RF power amplifier which preserves the incoming signal amplitude variation. If the transmitted signal is distorted by compression of the transmitter's power amplifier, then the receiver will be unable to accurately recovering the information encoded in the amplitude portion of the modulation.

2.5   Load and Source Pull Techniques and Impedance Matching of RF Power Amplifiers

The load and source pull techniques [17] developed by the electronics industry is a very powerful and efficient technique to experimentally optimize key performance metrics of an RF power amplifier. *Load pull involves varying the load impedance presented to a device under test (in this case, an RF power amplifier) and monitoring a single or set of performance parameters. So, if the selected performance metric to be monitored is an RF power amplifier (e.g., class A at some predefined operating frequency f_o) efficiency, the test engineer can determine that particular value of the load impedance for which the efficiency is as close as possible (within preset tolerances) to the theoretical limit of 50%. This implies that the load impedance value so determined provides the best impedance matching under that set of operating conditions.* However, in almost all real-world applications, the designer provides the source and load impedances and has to design and implement appropriate impedance matching sub-circuits for the input and output ports. *Load and/or source pull schemes enable the designer to assess what source and load impedance the amplifier expects to see at its input and output ports*. This will become clear when one examines the detailed design examples presented in the next chapter. The source pull technique is identical to the load pull technique, except that it is performed to optimize the source impedance. Both these schemes are flexible and may be performed together or independently of each other.

Traditionally, load pull analysis is used to construct a set of contours (sometimes on a Smith Chart). These contours determine the maximum power output to be achieved with a given load impedance. *These contours enable the engineer to assess the actual impedance a device should see when it is used in an amplifier.*

The experimental setup for load pull analysis is simple. The amplifier under test, referred to as "device under test" (DUT), has a load and source impedance at its two ports, and in between each of these impedances and the DUT, a tuner is placed. The input port tuner is used to ensure that the large signal input power is constant, even when the output impedance and the corresponding tuner are adjusted.

At start, both tuners are adjusted to obtain the maximum constant output power, which corresponds to the best possible match for the optimum output impedance and is the center of the loci of contours described earlier.

In the next sequence of steps, both the output impedances are changed and the input port tuner is adjusted to provide conjugate matching and constant input power. This is repeated as many times as possible and for each constant output power value, a set of loci are generated which provide the impedances which in turn provide that power. Each contour represents the maximum output power that can be obtained with a predefined load impedance. Identical exercise may be performed with the source impedance.

The load pull scheme can be analyzed in a straightforward manner. The load impedance for maximum linear power at the 1 dB compression point is $R_{OPT,LOAD} = \frac{V_{CC}}{I_{MAX}}$, where I_{MAX} is the maximum current. The case of current limited power occurs when the load impedance is less than $R_{OPT,\ LOAD}$. In this, more current is needed to generate a larger voltage swing and power. The reverse happens when the load impedance is larger than $R_{OPT,\ LOAD}$ and corresponds to voltage limited power, when a larger voltage swing is necessary to generate a larger current swing and thus power. For both cases, it is assumed that the input signal power is sufficiently high to drive the transistor to maximum current or voltage.

For current limited power case, with $Z_{LOAD} = R_{LOAD} + jX_{LOAD}$ the maximum linear power is $P_{OUT,CURRENT\ LIMITED} = \frac{I_{PEAK}^2 R_{LOAD}}{2}$ and the peak-peak output voltage is $\boxed{V}_{PP} \boxed{=} I_{MAX} \sqrt{R_{LOAD}^2 + X_{LOAD}^2}$. Using $R_{OPT,LOAD} = \frac{V_{CC}}{I_{MAX}}$, the maximum current is $I_{MAX} = \frac{2.0\,V_{CC}}{R_{OPT,LOAD}}$ and the peak-peak voltage can be expressed as $\boxed{V}_{PP} \boxed{=} \left(\frac{2.0\,V_{CC}}{R_{OPT,LOAD}}\right) \sqrt{R_{LOAD}^2 + X_{LOAD}^2}$. As the voltage swing can never exceed V_{CC}, for $\boxed{V}_{PP} \boxed{=} 2.0\,V_{CC}$ to hold. It must be that $R_{LOAD}^2 + X_{LOAD}^2 = R_{OPT,LOAD}^2$ or $\boxed{X}_{LOAD}^2 \boxed{+} R_{OPT,LOAD}^2 = R_L^2$.

The voltage limited case can be analyzed using admittances instead of impedances.

Detailed RF power amplifier design examples are presented and analyzed in the next chapter. The analysis of each of these design examples uses the SPICE [18–21] transient analysis feature, along with the core steps and principles of the load and source pull scheme. As both load and source pull are experimental schemes, they are very flexible.

2.6   Addressing Drawbacks of Narrow Band and Distributed RF Power Amplifiers

Each of the schematics (Figs. 2.2, 2.3a, b, 2.4a, b, 2.5, 2.6a, b, c, 2.7a, b, 2.8, and 2.9) of the RF power amplifiers analyzed so far represents narrow band RF power amplifiers that use discrete passive (capacitor, inductor, and resistor) elements. The key drawback of narrow band amplifiers is that as the gain-bandwidth product has to remain constant, as operating frequencies increase, the gain decreases. Distributed amplifiers are a very effective way to tackle this problem.

Fig. 2.10 Typical distributed amplifier configuration

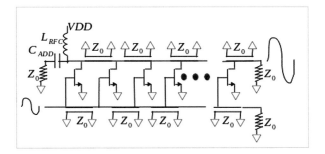

A distributed amplifier [27] uses a number of field effect transistors (FET – typically junction type, as compared to metal oxide field effect transistors) fed by a *periodic structure* at the input that resembles a terminated transmission line. The combination of FET capacitance with the high-impedance connection lines resembles a lumped-element version of a 50 ohm characteristic impedance transmission line, providing effective impedance match between the input and output ports. *This boosts bandwidth, often as high as 100 GHz, without altering the gain, resulting in a wide band constant (within applicable tolerances) gain RF power amplifier. The limitation is that some FETs get more power stressed than others in the chain.*

The operation of a distributed amplifier is understood by invoking the concept of the traveling-wave tube amplifier (TWTA). A distributed amplifier consists of a pair of transmission lines, each with a characteristic impedance of Z_0 independently connecting the gates and drains of the field effect transistors (FETs) (Fig. 2.10). An RF input signal is applied to one terminal of the section of transmission line connected to the gate of the first FET. As the input signal propagates down the input line, each individual FET responds to the forward traveling input step by inducing an amplified complementary forward traveling wave on the transmission line connected to the drains of each of the FETs. The key underlying assumption is that the delays of each of the input and output transmission lines can be made approximately equal by selecting their propagation constants and lengths. As a result the output signals from each transistor sum in phase.

The transconductance of each FET is g_m and the impedance seen by each FET is $\frac{Z_0}{2}$, and for a distributed amplifier with n stages, the overall voltage gain is $A_v = \frac{n g_m Z_0}{2}$. This expression for the gain is linear, and thus the total gain of a distributed amplifier is much higher than the unity gain of each individual stage. In practice, the number of stages is limited by the attenuation of the input signal as it propagates down the input transmission line. Bandwidth is limited by impedance mismatches due to frequency-dependent FET parasitics. Although the distributed amplifier boosts gain, its efficiency is mediocre.

References

1. Grebennikov A (2015) RF and microwave power amplifier design, 2nd edn. Mc-Graw Hill Educational. ISBN 978-07-0-182862-8
2. Walker J (ed) (2012) Handbook of RF and microwave power amplifier. Cambridge University Press. ISBN 978-0-521-76010-2
3. Bahl I. Fundamentals of RF and microwave tranistor amplifiers. John Wiley and Sons. ISBN 978-0-470-39166-2
4. Grebennikov A, Kumar N, Binboga SY (2017) Broadband RF and microwave amplifiers. CRC Press. ISBN 9781138800205 - CAT# K32788
5. Eroglu A (2015) Introduction to RF power amplifier design and simulation. CRC Press. ISBN 978-1-4822-3165-6
6. Cripps SC (2006) RF power amplifiers for wireless communications. Artech House. ISBN 10: 1-59693-018-7. ISBN 13: 978-1-59693-018-7
7. Cripps SC (2002) Advanced techniques in RF power amplifier design. Artech House Print on Demand. ISBN 10: 1580532829. ISBN 13: 9781580532822
8. Sokal's original class E amplifier paper from.: https://people.physics.anu.edu.au/~dxt103/160m/class_E_amplifier_design.pdf
9. Another source for Sokal's paper.: https://people.eecs.berkeley.edu/~culler/AIIT/papers/radio/Sokal%20AACD5-poweramps.pdf
10. Slade G. Notes on designing class E RF amplifiers. https://www.researchgate.net/publication/320623200_Notes_on_designing_Class_E_RF_power_amplifiAers
11. Albulet M (2001) RF power amplifiers. Noble Publishing. ISBN 1-884932- 12-6
12. Kazimierczuk MK (2015) RF power amplifiers, 2nd edn. John Wiley & Sons. Print ISBN:9781118844304. Online ISBN:9781118844373A
13. Shirvani A, Wooley BA. Design and control of RF power amplifiers. Springer. Ebook ISBN 978-1-4757-3754-7. Hardcover ISBN 978-1-4020-7562-9
14. Rudiakova AN, Krizhanovski V (2006) Advanced design techniques for RF power amplifiers. Springer. ISBN 978-1-4020-4639-1
15. Pozar DM. Microwave engineering, 4th edn. John Wiley and Sons Publication. ISBN 978-0-470-63155-3
16. Colantonio P, Giannini F, Limiti E. High efficiency RF and microwave solid state power amplifiers. John Wiley and Sons. Print ISBN:9780470513002. Online ISBN:9780470746547. https://doi.org/10.1002/9780470746547
17. Good introduction on load pull from.: http://mwrf.com/test-measurement/impedance-tuning-101
18. Latest Ngspice version 31 user guide and manual from.: http://ngspice.sourceforge.net/docs/ngspice-manual.pdf
19. LTSpice users guide and manual from.: https://ecee.colorado.edu/~mathys/ecen1400/pdf/scad3.pdf
20. Pspice users guide and manual from.: https://www.seas.upenn.edu/~jan/spice/PSpice_UserguideOrCAD.pdf
21. HSpice users guide and manual from.: https://cseweb.ucsd.edu/classes/wi10/cse241a/assign/hspice_sa.pdf
22. Ebook of the all-time classic C programming language book by the creators of the C computer language Brian Kernighan and Dennis Ritchie, can be downloaded easily from.: http://www2.cs.uregina.ca/~hilder/cs833/Other%20Reference%20Materials/The%20C%20Programming%20Language.pdf
23. https://www.reed.edu/physics/courses/Physics331.f08/pdf/Fourier.pdf
24. http://www.people.fas.harvard.edu/~djmorin/waves/Fourier.pdf

25. Doherty WH (1936) A new high efficiency power amplifier for modulated waves. Proc IRE 24:1163–1182
26. Markos AZ, Colantonio P, Giannini F, Giofrè R, Imbimbo M, Kompa G (October 2007) A 6W uneven Doherty power amplifier in GaN technology. In: Proceedings of the 2nd European microwave integrated circuits conference, Munich, Germany
27. Grebennikov A, Kumar N, Distributed RF (2015) Power amplifiers for RF and microwave communications. Artech House. ISBN: 9781608078318

Chapter 3
Radio Frequency Amplifier Design Examples and Performance Optimization with SPICE and Load and Source Pull Schemes

3.1 Introduction

This chapter contains an exhaustive set of design examples of electronic radio frequency (RF) power amplifier classes A, B, AB, C, D, E, F, G, and H, operating at 100 s of MHz (megahertz) to 10 s of GHz (Gigahertz). The design and analysis workflow are simple. Each RF amplifier class has its dedicated C computer language [7] executable that uses designer supplied amplifier circuit parameters (e.g., bi-junction transistor collector-emitter voltage and output load target power) to compute the amplifier circuit passive component (capacitor, inductor, resistor) values and arranges these results as a text SPICE [1–4] input format *netlist*. The text SPICE [1–4] input format *netlist* can be edited easily, allowing for design space exploration, specifically *the electronic industry standard load and source pull techniques*. The fresh generated netlist can be simulated and analyzed with any available open-source or proprietary SPICE [1–4] simulator, using its transient analysis feature. *The load and source pull techniques that are used with the design examples exploit the key concepts of these two schemes – load and source impedances that can be varied judiciously to maximize amplifier performance characteristics – in this case, amplifier efficiency. To demonstrate how easily the load and source pull methods can be incorporated into the SPICE [1–4] netlists, the load and source impedances are pure resistors. The more realistic case of complex load and source impedances will be demonstrated in the near future, in a future edition of this book.* The advantages of using SPICE [1–4] are as follows:

- Decades of enhancements and improvements to SPICE [1–4]'s built-in simulation engine has guaranteed that simulation results for any given circuit under test are *authentic, reliable, verifiable, and repeatable*. This has made SPICE [1–4] the gold-standard electrical/electronic circuit simulation tool in industry.
- A number of open-source (Ngspice, LTSpice, etc.,) and proprietary (HSpice, PSpice, etc.,) versions of SPICE [1–4] are available, allowing the user to use any

A. Banerjee, *Practical RF Amplifier Design and Performance Optimization with SPICE and Load- and Source-pull Techniques*, https://doi.org/10.1007/978-3-030-62512-2_3

available version. Most importantly, results generated by one SPICE [1–4] can be verified by another, making the results reliable and authentic. All design examples in this book have been simulated with the latest version of Ngspice, version 31.

- Almost all SPICE [1–4] simulators allow text SPICE input format *netlists. These can be easily edited/modified for design space exploration.*
- SPICE [1–4] semiconductor device models (to be used by a given SPICE simulation engine), in this case radio frequency (RF) bi-junction and field effect transistors, are freely downloadable from reputed semiconductor device manufacturer's Web sites [5, 6].
- Transient analysis feature of any SPICE [1–4] simulation engine exploits large-signal steady-state circuit behavior model and thus the designer does not need to tackle circuit start-up time, small-signal S (scattering) parameters supplied by semiconductor device manufacturers. The key large-signal semiconductor device models used in SPICE [1–4] transient analysis are supplied by the manufacturers of these device [5, 6]. This is a significant advantage, as extracting large-signal S parameters requires either high-end CAD (Computer Aided Design) tools or expensive test equipment as spectrum analyzers.

The SPICE [1–4] transient analysis output is automatically written into another text file, which is analyzed by another C computer language [7] executable to extract statistical information (e.g., root mean square – RMS) current and voltage. Armed with this knowledge, the analysis of various RF amplifiers start. In all subsequent sections of this chapter, computer-generated text is displayed in bold font. All design examples use the RF bi-junction transistor pair BFR92A (NPN) [5] and BFT92 (PNP) [6], unless otherwise mentioned. *It must be noted that the load and source pull schemes are used to optimize the key RF amplifier performance metric efficiency in the subsequent design examples. Similar exercises can be performed for the other amplifier performance metrics such as power-added efficiency, gain, etc.,*

3.2 500 MHz Class A Amplifier: Design with SPICE and Efficiency Boost from 16% to 40% with Load Pull Scheme

The C computer language [7] executable ***amplifierrfclsa*** accepts the minimum bi-junction transistor beta (DC forward current gain) value (h_{FE}), the collector current (milliampere), the target load output power (watts), bi-junction transistor collector-emitter voltage (V_{CE}), and operating frequency (MHz). It computes the values for the input/output DC blocking capacitors, emitter bypass capacitor and base bias, collector and emitter resistors, and class A RF power amplifiers (Chap. 2, Sect. 2.3.1). Most importantly, ***amplifierrfclsa*** arranges these computed values in a text SPICE [1–4] input format *netlist* that can be used with any available SPICE [1–

4] simulator. **amplifierrfclsa** is invoked from the command line and help information is displayed when **./amplifierrfclsa** is typed at the Linux shell command prompt.

$ ./amplifierrfclsa
incorrect|insufficient input
interactive mode
./amplifierrfclsa i|I
batch - command line argument mode
./amplifierrfclsa b|B|c|C
<minimum transistor beta>
<maximum collector current(mA)>
<target output power(Watts)>
<collector emitter voltage(Vce - V)>
<operating frequency(MHz)>
check transistor datasheet for parameter values
sample batch command line input - RF BJT BFR92A
./amplifierrfclsa b|B|c|C 65 14 5 10 500

Using the sample command line input generates a text SPICE netlist amprfclsa.cir that is used by the Ngspice 31 simulator.

SPICE netlist amprfclsa.cir is as follows:

$ ./amplifierrfclsa C 65 14 5 10 500
SPICE netlist amprfclsa.cir

The contents of the generated SPICE *netlist* are as follows:

.INCLUDE hiperftransistormodels

**** CLASS A AMPLIFIER WITH BASE BIAS**

**** INPUT|OUTPUT IMPEDANCE MATCHING MUST BE ADDED - LOAD SOURCE PULL**

.PARAMS VCC=25.000 RC=892.857 RE=178.571 RB0=11607.143
+ RB1=1703.801
.PARAMS CDCBLK=2.000000e-09 CEBYP=1.783439e-13 TS=1.000000e-10
+ RL=62.500 R=50.0
.PARAMS FREQ=5.000000e+08 AMPL=18.750 C0=1.0e-3 LRFC=1.0E-6

**** INITIAL ESTIMATE RL=62.500 R=50.0**

C0 3 0 {C0}
CDCBLK0 7 4 {CDCBLK}
CDCBLK1 5 8 {CDCBLK}
CEBYP 6 0 {CEBYP}
LRFC 3 11 {LRFC}

```
RB1 11 4 {RB0}
RB2 4 0 {RB1}
RC 11 5 {RC}
RE 6 0 {RE}
RS 2 7 {R}
RL 10 0 {RL}
XQ0 5 9 6 BFR92A

VCC 1 0 DC {VCC} AC 0.0
VSIG 2 0 DC 0.002
+ SIN(0 {AMPL} {FREQ} 0 0 0)
VTST0 1 3 DC 0.0 AC 0.0
VTST1 4 9 DC 0.0 AC 0.0
VTST2 8 10 DC 0.0 AC 0.0

.OPTIONS METHOD=GEAR NOPAGE RELTOL=1m
.IC
.TRAN {TS} 50.0us 30.0us UIC
** OUTPUT
.PRINT TRAN V(10) I(VTST2)
** INPUT
.PRINT TRAN V(9) I(VTST1)
** DC SUPPLY CURRENT
.PRINT TRAN I(VTST0)
.END
```

It is clear that load and source pull schemes can be integrated into this SPICE netlist simply by modifying the load and/or source impedances and measuring the amplifier efficiency for each combination of load and source impedances. This process is iteratively repeated till the measured amplifier efficiency achieves or exceeds a predefined threshold value. The Ngspice 31 simulator is invoked from the shell command prompt as: **ngspice -b amprfclsa.cir > out**

The Ngspice 31 simulator writes the transient analysis output to a simple text file named **out**. This text file serves as the input for a supplied utility C computer language program **rmscalc**, which computes the RMS (root mean square) current and voltage.

For a class A amplifier, the maximum theoretical efficiency is 50%, so a reasonable target efficiency value would be approximately 45% or more. The results of the load and source pull scheme, as applied to the above class A amplifier *netlist*, are listed in Table 3.1.

Careful examination of data in the above table indicates the following (keeping in mind that the theoretical maximum efficiency of a class A amplifier is 50%):

- By adjusting the load and source impedances, the efficiency of this class A RF amplifier can be boosted from 16% to 40%.
- To operate at 500 MHz with 40% efficiency, the source and load impedances must be 25.0 and 500 Oms, respectively. Since in a real-world design and application case it is impossible to change the load and source impedances, two

Table 3.1 Load source pull current, voltage, and efficiency for 500 MHz class A amplifier

Load impedance 62.5 ohm			Source impedance 50.0 ohm			
RMS load current (A)	RMS load voltage (V)	RMS load power (watts)	RMS power supply current (A)	RMS power supply voltage (V)	RMS supplied power (watts)	Efficiency
0.035	2.217	0.077595	0.018684	25.0	0.475	0.1633
Load impedance 125.0 ohm			Source impedance 50.0 ohm			
0.031	3.865	0.119753	0.03604	25.0	0.4505	0.2658
Load impedance 220.0 ohm			Source impedance 40.0 ohm			
0.026	5.655	0.14707	0.017404	25.0	0.4451	0.3304
Load impedance 250.0 ohm			Source impedance 40.0 ohm			
0.025	6.212	0.1553	0.017198	25.0	0.42945	0.3646
Load impedance 250.0 ohm			Source impedance 30.0 ohm			
0.026	6.482	0.170532	0.01718	25.0	0.4295	0.3970
Load impedance 500.0 ohm			Source impedance 25.0 ohm			
0.018	8.885	0.15993	0.0159	25.0	0.3975	0.4023

simple L impedance matching transformers must be designed for the input and output ports of the amplifier. This is left as an exercise for the reader.

- Further adjustment of the load and source impedances can boost the efficiency closer to the theoretical limit of 50%. This is left as an exercise for the reader.

It is noteworthy that the class A RF power amplifier lacks a resonator, as compared to class B, C, D, E, and F RF power amplifiers.

3.3 750 MHz Class B Single-Sided Amplifier: Design with SPICE and Efficiency Boost from 40% to 71% with Load Pull

The supplied C computer language [7] executable ***amplifierrfclsb*** is used to generate the SPICE [1–4] netlist for the single-sided class B RF amplifier (Fig. 2.3a). Although the class B can also exist as a double-sided version, for RF applications, the single-sided version is used. ***amplifierrfclsb*** requires three amplifier circuit parameters from the designer – target load output power, bi-junction transistor collector-emitter current, and operating frequency. It computes values for the amplifier circuit's input transformer's primary, secondary inductors, the output DC blocking capacitor, and the resonator capacitor, inductor product. Finally, it arranges all the computed values as a text SPICE [1–4] input format netlist. The values of the resonator capacitor and inductor must be assigned manually by the designer in the generated SPICE [1–4] netlist. The reason for this manual assignment of the resonator capacitor, inductor values will be explained next. ***amplifierrfclsb*** is invoked from the command line, and typing *./amplifierrfclsb* from the command line generates help information.

The following command line argument list is used to generate the SPICE netlist for a 750 MHz single-sided RF class B amplifier with the target load output power of 10.0 watts and collector-emitter voltage of 12.5 volts.

./amplifierrfclsb C 750 10 12.5 ss

The generated text SPICE netlist named **amprfclsb.cir** is listed below.

.INCLUDE hiperftransistormodels

** IMPEDANCE MATCHING AT START, END MUST BE ADDED LOAD, SOUTCE PULL

** RESOMATOR C, L VALUES TO BE
** ESTIMATED FROM LC PRODUCT
.PARAMS CDCBLK=1.333333e-09 LRFC=1.061571e-08
RL=7.812
.PARAMS LT=1.061571e-09 TS=6.666667e-11
LCPROD=4.507733e-20
.PARAMS FREQ=7.500000e+08 VCC=12.500 R=50.000 AMPL=7.5 CR=1.0e-15
LR=4.507733e-5

** INITIAL RSTIMATE RL=7.812

** CLASS B RF AMPLIFIER SINGLE SIDED

C0 3 0 1.0e-6
CDCBLK0 6 7 {CDCBLK}
CR 7 0 {CR}
LRFC 3 6 {LRFC}
** INPUT TRANSFORMER
LT0 2 0 {LT}
LT1 5 0 {LT}
k0 LT0 LT1 0.99
LR 7 0 {LR}
RL 8 0 {RL}
XQ0 6 10 0 BFR92A

VCC 1 0 DC {VCC} AC 0.0
VSIG 2 0 DC 0.0 SIN(0 {AMPL} {FREQ} 0 0 0)

VTST0 1 3 DC 0.0 AC 0.0
VTST1 5 10 DC 0.0 AC 0.0
VTST2 7 8 DC 0.0 AC 0.0

.OPTIONS METHOD=GEAR NOPAGE RELTOL=1m
.IC

```
.TRAN {TS} 100.0us 80.0us UIC
** OUTPUT
*.PRINT TRAN V(8) I(VTST2)
** INPUT
*.PRINT TRAN V(10) I(VTST1)
**DC SUPPLY CURRENT
.PRINT TRAN I(VTST0)
.END
```

Resonator C, L Value Selection amplifierrfclsb computes the resonator LC product, not the individual C, L values. These two values need to be added manually by
the designer. From the above SPICE [1–4] netlist, one terminal of the DC blocking
capacitor is connected to the bi-junction transistor collector terminal, and the other
capacitor terminal is connected to common node shared by the resonator capacitor,
resonator inductor, and the load. **Since the signal arriving at this common node is
purely AC (zero DC component), the resonator capacitor and inductor values
must be selected such that they offer maximum possible AC resistance (capacitor, inductor reactance) at the operating frequency, thereby ensuring that the
maximum output signal power is delivered to the load. The capacitive reactance
at 750 MHz is 212314.225 ohm, the inductive reactance is 212314.2243 ohm,
and the product of the capacitor and inductor values is 4.507733e-20.** *The
resonator is key to the single-sided class B amplifier's operation. The NPN
bi-junction can switch on only during the time interval during which the amplifier
input signal (applied to the transistor base) has positive polarity (positive half
cycle). During the half cycle period that the transistor is switched off, the resonator
compensates, and the output, appearing at the load, is fully sinusoidal.*

*The generated SPICE [1–4] netlist is used by the Ngspice 31 simulator, and load
pull is easily integrated into the analysis simply by altering the load impedance and
executing the Ngspice simulator with the new load impedance value. For each new
value of the load impedance, the RMS load output power and supplied DC power
are computed, and the efficiency is computed.* The results obtained from the load pull
experiment are listed in Table 3.2.

Table 3.2 Load pull current, voltage, and efficiency for 750 MHz single-sided class B amplifier

Load impedance 7.812 ohm			Source impedance 4.71 ohm			
RMS load current (A)	RMS load voltage (V)	RMS load power (watts)	RMS power supply current (A)	RMS power supply voltage (V)	RMS supplied power (watts)	Efficiency
0.45	3.512	1.5804	0.476	12.5	5.95	0.4001
Load impedance 15.624 ohm			Source impedance 4.71 ohm			
0.388	6.232	2.486568	0.433	12.5	5.4125	0.4594
Load impedance 31.25 ohm			Source impedance 4.71 ohm			
0.326	10.178	3.318028	0.375	12.5	4.6875	0.7078

Clearly, the final efficiency value calculated is very close to the class B theoretical maximum efficiency of 78.75%.

3.4   800 MHz Class AB RF Amplifier: Design with SPICE and Efficiency Boost from 16% to 75% with Load Pull

The supplied C computer language [7] executable ***amplifierrfclsab*** generates the text SPICE [1–4] input format netlist for a class AB RF amplifier. Since the class AB amplifier contains an NPN RF transistor, and its complementary RF PNP transistor, the designer for both the NPN and PNP transistors has to supply the minimum beta (h_{FE}) values, the collector currents, and the collector-emitter voltage (V_{CE}). The designer also has to supply the operating frequency (MHz) and target load output power (watts). The supplied ANSI C computer language executable ***amplifierrfclsab*** computes values for the collector resistors for both transistor types, collector resistors for both the transistor types, load impedances for both transistor types, the values for the DC blocking capacitor. The actual load impedance used for the SPICE [1–4] simulation is minimum of these two computed load impedance values. The computed values are all arranged as a text SPICE input format netlist. The RF NPN transistor used is BFR92A [5], and its complementary PNP transistor is BFT92 [6]. Typing *./amplifierrfclsab* at the command prompt generates the help information. The executable is best invoked in the batch (i.e., using the argument list)

```
$ ./amplifierrfclsab
incorrect|insufficient arguments
interactive mode
./amplifierrfclsab i|I
batch|command line argument mode
./amplifierrfclsab b|B|c|C
<mimimum NPN transistor beta>
<minimum PNP transistor beta>
<NPN collector current(mA)>
<PNP collector current(mA)
<operating frequency(MHz)>
<target load power(Watts)
<NPN collector - emitter voltage(V)>
<PNP collector - emitter voltage(V)>
sample command line input BFR92A(NPN) and BFT92(PNP)
./amplifierrfclsab b 65 20 14 -14 600 10 10 -10
```

So the sample command line input with the operating frequency modified to 800 MHz yields the text SPICE netlist.

```
.INCLUDE hiperftransistormodels
.INCLUDE diodemodelsgen
```

** IMPEDANCE MATCHING AT START, END MUST BE ADDED - LOAD, SOURCE
PULL

.PARAMS VCCP=12.500 RBNPN=58035.714
RCNPN=892.857
.PARAMS VCCM=-12.500 RBPNP=17857.143
RCPNP=892.857
.PARAMS FREQ=8.000000e+08 CDCBLK=1.250000e-09 TS=6.250000e-11
RLOAD=31.248 R=10.0
AMPL={0.75*VCCP}

** INITIAL ESTIMATE RLOAD=7.812 R=50.0

CDCBLK0 14 4 {CDCBLK}
CDCBLK1 9 10 {CDCBLK}
D0 5 4 1N4007
D1 4 6 1N4007
RBIASNPN 12 5 {RBNPN}
RBIASPNP 13 6 {RBPNP}
RCNPN 12 7 {RCNPN}
RCPNP 13 8 {RCPNP}
RLOAD 11 0 {RLOAD}
RS 3 14 {R}
XQ0 7 5 9 BFR92A
XQ1 8 6 9 BFT92

VCCP 1 0 DC {VCCP} AC 0.0
VCCM 2 0 DC {VCCM} AC 0.0
VSIG 3 0 DC 0.0001 SIN(0 {AMPL} {FREQ} 0 0 0)
VTST0 1 12 DC 0.0 AC 0.0
VTST1 2 13 DC 0.0 AC 0.0
VTST2 10 11 DC 0.0 AC 0.0

.OPTIONS METHOD=GEAR NOPAGE RELTOL=1m
.IC
.TRAN {TS} 100.0us 70.0us UIC
** OUTPUT
*.PRINT TRAN V(11) I(VTST2)
** +VE DC CURRENT
.PRINT TRAN I(VTST0)
** -VE DC CURRENT
*.PRINT TRAN I(VTST1)
.END

As the class AB amplifier does not use any resonator, there is no need to select the corresponding capacitor and inductor values. Also, as both NPN and PNP RF transistors are used in the circuit, positive and negative polarity power supplies are required. Load and/or source pull schemes are easily integrated into the SPICE [1–4] netlist simply by altering the load and/or source impedances and by executing the simulation with the modified netlist iteratively. The results of the load source pull exercise are given in Table 3.3. **Note that the supplied RMS power value used in**

Table 3.3 Load and source pull RMS load power, RMS DC source power, and efficiency for class AB amplifier at 800 MHz

Load impedance 7.812 ohm			Source impedance 50.0 ohm			
RMS load current (A)	RMS load voltage (V)	RMS load power (watts)	RMS power supply current (A)	RMS power supply voltage (V)	RMS average supplied power (watts)	Efficiency
0.056	0.440	0.02464	0.012	12.5	0.15	0.1643
			0.013	12.5	0.15	
Load impedance 15.624 ohm			Source impedance 50.0 ohm			
0.048	1.489	0.071472	0.012	12.5	0.15	0.4765
			0.012	12.5	0.15	
Load impedance 15.624 ohm			Source impedance 30.0 ohm			
0.053	1.651	0.087503	0.012	12.5	0.15	0.5833
			0.012	12.5	0.15	
Load impedance 15.624 ohm			Source impedance 20.0 ohm			
0.057	1.772	0.101004	0.012	12.5	0.15	0.6734
			0.012	12.5	0.15	
Load impedance 15.624 ohm			Source impedance 10.0 ohm			
0.060	1.873	0.11238	0.012	12.5	0.15	0.7492
			0.012	12.5	0.15	

the final efficiency calculation is an average of the supplied RMS power value measured with respect to the positive polarity power supply and the supplied RMS power measured with respect to the negative polarity power supply.

Clearly, the efficiency has been boosted from 16% to 75%. *As the PNP bi-junction transistor switches on when the NPN bi-junction transistor is switched off, efficiency is increased.*

3.5   750 MHz Zero-Base Bias Class C Amplifier: Design with SPICE and Efficiency Boost from 34% to 89% with Load Pull

The supplied C computer language [7] executable *amplifierrfclsc* can generate the SPICE input text netlist for both a zero-base bias and a fixed-base bias class C amplifier. The zero-base bias class C amplifier is very similar to a single-sided class B amplifier. Double-sided class C amplifiers do not exist. *amplifierrfclsc* is typically used in the batch|command line argument mode with command line argument list.

Typing ./**amplifierrfclsc** at the command prompt generates the help information as listed below:

$ ./amplifierrfclsc

incorrect|insufficient arguments
interactive mode
./amplifierrfclsc i|I
batch|command line argument mode
./amplifierrfclsc b|B|c|C
<operating frequency(MHz)>
<target load power(Watts)>
<DC supply voltage(V) -- check transistor datasheet>
<zero z|Z OR fixed base bias f|F>

sample batch|command line input - BJT BFR92A zero base bias

./amplifierrfclsc b 500 5 13 z|Z
sample batch|command line input fixed base bias
./amplifierrfclsc C 750 5 12.5 f|F

Using the sample command line input for zero-base bias class C amplifier with the operating frequency modified to 800 MHz generates the text SPICE netlist listed below:

.INCLUDE hiperftransistormodels

**** IMPEDANCE MATCHING AT START, END TO BE ADDED LOAD, SOURCE PULL**

**** RESONATOR C, L VALUES TO BE**
**** COMPUTED FROM LC PRODUCT VALUE**

.PARAMS FREQ=7.500000e+08 LRFC=2.123142e-08
CDCBLK=1.333333e-09 LCPROD=4.507733e-20
TS=6.666667e-11
.PARAMS VCC=13.000 RL=67.8
+ AMPL=7.5 CR=1.0e-15 LR=4.507733e-5 R=10.0
.PARAMS FLLIM=7.500000e+07 FHLIM=1.500000e+09
TOT=25000

**** INITIAL ESTIMATE RL=16.900 RS=50.0**

**** ZERO BASE BIAS**

CDCBLK 4 5 {CDCBLK}
CR 5 0 {CR}
LR 5 0 {LR}
LRFC0 3 4 {LRFC}
LRFC1 11 0 {LRFC}
RL 6 0 {RL}
RS 2 11 {R}
XQ0 4 11 0 BFR92A

```
VCC 1 0 DC {VCC} AC 0.0
** TRANSIENT ANALYSIS
VSIG 2 0 DC 0.0001 SIN(0 {AMPL} {FREQ} 0 0 0)
** AC(SMALL SIGNAL) ANALYSIS
**VSIG 2 0 DC 0.0001 AC {AMPL}
VTST0 1 3 DC 0.0 AC 0.0
VTST1 5 6 DC 0.0 AC 0.0

.OPTIONS METHOD=GEAR NOPAGE RELTOL=1m
** TRANSIENT ANALYSIS
.IC
.TRAN {TS} 90.0us 70.0us UIC
** OUTPUT
*.PRINT TRAN V(6) I(VTST1)
** DC INPUT CURRENT
.PRINT TRAN I(VTST0)
** AC(SMALL SIGNAL) ANALYSIS
*.AC LIN {TOT} {FLLIM} {FHLIM}
*.PRINT AC V(6)
.END
```

The resonator C, L values are selected using the same procedure as that used in the case of the single-sided class B amplifier (Sect. 3.3). The main goal of this procedure to estimate the resonator's capacitor, inductor (C, L) values is to maximize capacitive and inductive reactances. This in turn minimizes AC signal current flowing through the resonator capacitor to ground. The inductor acts as a high resistance at high frequencies, so that a high inductance value is a large value resustor at RF frequencies. As before, load and/or source pull schemes are easily integrated into the SPICE netlist by adjusting the source and load impedances. The results are listed in Table 3.4.

The class C amplifier requires a resonator to compensate for the input signal half cycle period during which the signal polarity is negative and the NPN bi-junction is switched off. The values for the resonator capacitor and inductor are selected using similar reasoning as that used for selecting the capacitor inductor values for the single-sided class B amplifier. *amplifierrfclsc* can also be used to design and then analyze a class C *fixed-base bias* amplifier. This is left as an exercise for the reader.

Table 3.4 Load and source pull RMS load power, RMS DC source power, and efficiency for class AB amplifier at 750 MHz

Load impedance 16.9 ohm			Source impedance 20.0 ohm			
RMS load current (A)	RNS load voltage (V)	RMS load power (watts)	RMS power supply current (A)	RMS power supply voltage (V)	RMS supplied power (watts)	Efficiency
0.281	4.754	1.335874	0.31	12.5	3.875	0.3447
Load impedance 33.8 ohm			Source impedance 20.0 ohm			
0.234	7.912	1.850418	0.266	12.5	3.325	0.5538
Load impedance 67.6 ohm			Source impedance 20.0 ohm			
0.187	12.704	2.375648	0.225	12.5	2.8125	0.8446
Load impedance 67.6 ohm			Source impedance 10.0 ohm			
0.202	13.701	2.767602	0.249	12.5	3.1125	0.8892

3.6   500 MHz Double-Sided Class D Transformer-Controlled Voltage Source (TCVS) Amplifier: Design with SPICE Efficiency Boost from 21% to 75% with Load Pull

The supplied C computer language [7] executable *amplifierrfclsd* creates the text SPICE [1–4]input format netlist for all three types of class D amplifiers – single-sided, double-sided transformer-controlled voltage source (TCVS), and double-sided transformer-controlled current source (TCVS). To create the netlist for a TCVS class D amplifier, *amplifierrfclsd* requires three input parameters – operating frequency in MHz, target load output power, and transistor collector-emitter voltage. *amplifierrfclsd* is invoked from the command line with appropriate command line arguments, and typing *./amplifierrfclsd* at the shell command prompt yields the following help information:

```
$ ./amplifierrfclsd
insufficient|incorrect arguments
interactive mode
./amplifierrfclsd i|I
batch|command line mode
./amplifiweedclsd b|B|c|C
<operating frequency(MHz)>
<target load power(Watts)>
<collector emitter voltage(v)>
single sided(ss|sS|Ss|SS) or double sided(ds|dS|Ds|DS)
sample batch|command line input
single sided - BJT BFR92A
./amplifierrfclsd b|B|c|C 500 10 12.5 ss
double sided - transformer controlled voltage source
./amplifierrfclsd b|B|c|C 500 10 12.5 tcvs|TCVS
double sided - transformer controlled current source
./amplifierrfclsd b|B|c|C 500 10 12.5 tccs|TCCS
```

Using the sample command line argument for the TCVS class D amplifier generates a text SPICE [1–4] input format netlist named *amprfclsd.cir*, whose contents are displayed below:

```
.INCLUDE hiperftransistormodels

** IMPEDANCE MATCHING AT START, END MUST BE ADDED - LOAD, SOURCE
PULL
** RESONATOR C, L VALUES TO ESTIMATED
** FROM LC PRODUCT VALUE BELOW

.PARAMS CDCBLK=2.000000e-10 LCPROD=1.014240e-19
+ LRFC=1.592357e-08 RLOAD=7.812 TS=1.000000e-10
```

```
.PARAMS FREQ=5.000000e+08 VCC=12.500 AMPL={0.75*VCC} CR=1.0e-
8 LR=1.014240e-11 LT=1.0e-9

** INITIAL ESTIMATE RLOAD=7.812

** CLASS D TCVS

CR 8 9 {CR}
LR 9 10 {LR}
** INPUT SIGNAL XFRNR
LT0 2 0 {LT}
LT1 0 6 {LT}
LT2 7 0 {LT}
k0 LT0 LT1 0.99
k1 LT0 LT2 0.99
k2 LT1 LT2 0.99
** OUTPUT XFRMR
LT3 4 3 1.0e-6
LT4 3 5 1.0e-6
LT5 8 0 1.0e-6
k3 LT3 LT4 0.99
k4 LT3 LT5 0.99
k5 LT4 LT5 0.99
RL 11 0 {RLOAD}
XQ0 4 6 0 BFR92A
XQ1 5 7 0 BFR92A
VCC 1 0 DC {VCC} AC 0.0
** TRANSIENT ANALYSIS -- SIGNAL SOURCE
VSIG 2 0 DC 0.0001 SIN(0 {AMPL} {FREQ} 0 0 0)
VTST0 1 3 DC 0.0 AC 0.0
VTST1 10 11 DC 0.0 AC 0.0

.OPTIONS METHOD=GEAR NOPAGE RELTOL=1m
.IC
.TRAN {TS} 80.0us 60.0us UIC
** OUTPUT
*.PRINT TRAN V(11) I(VTST1)
** DC SUPPLY CURRENT
.PRINT TRAN I(VTST0)
,END
```

Like the previous design examples, both the load and source pull schemes can easily be incorporated into the above SPICE [1–4] netlist, and then the efficiency can be improved by iteratively simulating the above netlist with Ngspice 31. The load source pull results are listed in Table 3.5.

The reader is encouraged to explore ways to boost the efficiency of this amplifier even higher than 75%.

Table 3.5 Load and source pull impedance and efficiency for class D TCVS 500 MHz amplifier

Load impedance 7.812 ohm			Source impedance 3.14 ohm			
RMS load current (A)	RMS load voltage (V)	RMS load power (watts)	RMS power supply current (A)	RMS power supply voltage (V)	RMS supplied power (watts)	Efficiency
0.393	3.07	1.20651	0.447	12.5	5.5879	0.2159
Load impedance 15.624 ohm			Source impedance 3.14 ohm			
0.393	6.139	2.412627	0.433	12.5	5.4125	0.4457
Load impedance 31.248 ohm			Source impedance 3.14 ohm			
0.353	11.023	3.891119	0.465	12.5	5.8125	0.6694
Load impedance 64.5 ohm			Source impedance 3.14 ohm			
0.260	16.235	4.2211	0.447	12.5	5.5879	0.7554

3.7   750 MHz Double-Sided Class D Transformer-Controlled Current Source Amplifier: Efficient Boost from 16% to 61% with Load Pull

The supplied C computer language [7] executable ***amplifierrfclsd*** can be used easily with appropriate command line argument list to generate the text SPICE [1–4] netlist for a double-sided class D transformer-controlled current source type amplifier. The SPICE *netlist* is listed below.

.INCLUDE hiperftransistormodels

** IMPEDANCE MATCHING AT START, END MUST BE ADDED - LOAD, SOURCE PULL

** RESONATOR C, L VALUES TO ESTIMATED
** FROM LC PRODUCT VALUE BELOW

.PARAMS CDCBLK=1.333333e-10 LCPROD=4.507733e-20
+ LRFC=1.061571e-08 RLOAD=90.0 TS=6.666667e-11
.PARAMS FREQ=7.500000e+08 VCC=12.500
+ AMPL={0.75*VCC} CR=1.0e-15 LR=4.507733e-5
LT=1.0e-9
** INITIAL ESTIMATE RLOAD=7.812

** CLASS D TCCS
CR 8 0 {CR}
LR 8 0 {LR}
** INPUT SIGNAL XFRNR
LT0 2 0 {LT}
LT1 0 6 {LT}
LT2 7 0 {LT}
k0 LT0 LT1 0.99

```
k1 LT0 LT2 0.99
k2 LT1 LT2 0.99
** XFRMR
LT3 4 3 1.0e-6
LT4 3 5 1.0e-6
LT5 8 0 1.0e-6
k3 LT3 LT4 0.99
k4 LT3 LT5 0.99
k5 LT4 LT5 0.99
RL 9 0 {RLOAD}
XQ0 4 6 0 BFR92A
XQ1 5 7 0 BFR92A
VCC 1 0 DC {VCC} AC 0.0
** TRANSIENT ANALYSIS -- SIGNAL SOURCE
VSIG 2 0 DC 0.0001
 SIN(0 {AMPL} {FREQ} 0 0 0)
** AC(SMALL SIGNAL) ANALYSIS
**VSIG 2 0 DC 0.0001 AC {AMPL}
VTST0 1 3 DC 0.0 AC 0.0
VTST1 8 9 DC 0.0 AC 0.0

.OPTIONS METHOD=GEAR NOPAGE RELTOL=1m
** TRANSIENT ANALYSIS
.IC
.TRAN {TS} 80.0us 60.0us UIC
** OUTPUT
*.PRINT TRAN V(9) I(VTST1)
** DC SUPPLY CURRENT
.PRINT TRAN I(VTST0)
.END
```

As in the previous design examples, the load and source pull schemes are easily integrated into this SPICE [1–4] netlist by modifying the load and/or source impedances and by iteratively using the Ngspice 31 [1] simulator to measure and boost the efficiency. The results of the load and source pull experiments are listed in Table 3.6.

Clearly the efficiency of this class D amplifier is less than that of the previous class D amplifier. The reader is encouraged to determine the reason behind this.

3.8   500 MHz Class F Amplifier with Harmonic Tuning Up to Third Harmonic: Design with SPICE and Efficiency Boost from 35% to 80%

The supplied C computer language [7] executable *amplifierrfclsf* can generate text SPICE input format netlist with third, fifth, and seventh harmonic tuning. As mentioned earlier, a class F amplifier is an extension of the class B amplifier with harmonic tuning. So, there are two sets of resonators, one for tuning and the other, as in the case of the class B amplifier, for the fundamental, and *amplifierrfclsf* generates

Table 3.6 Load and source pull and efficiency for class D transformer-controlled current source (TCCS) amplifier at 750 MHz

Load impedance 7.812 ohm			Source impedance 4.71 ohm			
RMS load current (A)	RMS load voltage (V)	RMS load power (watts)	RMS power supply current (A)	RMS power supply voltage (V)	RMS supplied power (watts)	Efficiency
0.268	2.096	0.567128	0.288	12.5	3.6	0.156
Load impedance 15.624 ohm			Source impedance 4.71 ohm			
0.268	4.188	1.122384	0.299	12.5	3.7375	0.3003
Load impedance 31.248 ohm			Source impedance 4.71 ohm			
0.257	8.045	2.067565	0.327	12.5	4.0875	0.5058
Load impedance 62.496 ohm			Source impedance 4.71 ohm			
0.214	13.348	2.79638	0.376	12.5	4.7	0.6120

two sets of resonator capacitor, inductor product values. The designer has to manually assign the individual capacitor and inductor values, *keeping in mind that the amplifier output signal, just after the DC blocking capacitor, is purely AC without any DC offset. So, the capacitor-inductor values for the third harmonic are selected to provide minimum AC resistance, and correspondingly, the capacitor-inductor values for the resonator for the fundamental frequency are chosen to provide maximum AC resistance.*

Therefore, by invoking *amplifierrfclsf* using the command line argument list *./* **amplifierrfclsf b ss 500 10 12.5 3,** a text SPICE [1–4] input netlist named *amprfclsf. cir* is generated, which is listed below:

.INCLUDE hiperftransistormodels

** IMPEDANCE MATCHING AT START, END MUST BE ADDED - LOAD, SOUTCE PULL

** CLASS F RF AMPLIFIER WITH 3RD
** HARMONIC TUNING SINGLE SIDED

** RESONATOR C, L TO BE ESTIMATED
** FROM LC PRODUCT VALUE IN PARAMETER LIST

** IMPEDANCE MATCHING TO 50.0 OHM
** MUST BE ADDED SEPARATELY

** RESONATOR C, L FOR 3rd, 5th
** HARMONICS TO BE ESTIMATED
** FROM LC PRODUCT FOR VALUE
** FUNDAMENTAL - 3RD HARMONIC FREQUENCY
** IS 3xFUNDAMENTAL ...

```
.PARAMS CDCBLK=6.666667e-10 LRFC=3.184713e-08
+ LCPROD=1.014240e-19 LCPROD3=1.126933e-20

.PARAMS TS=3.333333e-11 FREQ=5.000000e+08 RL=7.812
.PARAMS CR=1.0e-14 LR=1.014240e-5
+ CR3=1.0e-8 LR3=1.126933e-12
+ AMPL=7.5 LT=1.0e-9 R0=50.0 VCC=12.5

** PULSED INPUT

.PARAMS PER=2.000000e-09 PW=1.000000e-09 TR=1.000000e-10
+ DLY=1.000000e-09

** INITIAL ESTIMATED LOAD IMPEDANCE 7.812 RS=3.16

CR 10 0 {CR}
CR3 9 10 {CR3}
LRFC 6 7 {LRFC}
LR 10 0 {LR}
LR3 9 10 {LR3}
** INPUT TRANSFORMER
LT0 2 0 {LT}
LT1 4 0 {LT}
k0 LT0 LT1 0.99
RL 11 0 {RL}
XQ0 7 8 0 BFR92A
VCC 1 0 DC {VCC} AC 0.0
VSIGP 3 0 DC 0.001 PULSE(0 {AMPL} 0 {TR} {TR} {PW} {PER})
VSIGM 2 3 DC 0.001 PULSE(0 {-AMPL} {DLY} {TR} {TR} {PW} {PER})
VTST0 1 6 DC 0.0 AC 0.0
VTST1 4 8 DC 0.0 AC 0.0
VTST2 10 11 DC 0.0 AC 0.0

.OPTIONS METHOD=GEAR NOPAGE RELTOL=1m
.IC
.TRAN {TS} 70.0us 50.0us UIC
** OUTPUT
*.PRINT TRAN V(11) I(VTST2)
** INPUT
*.PRINT TRAN V(8) I(VTST1)
** DC SUPPLY CURRENT
.PRINT TRAN I(VTST0)
.END
```

As in the previous design examples, the above SPICE netlist is the input for the Ngspice 31 simulator invoked from the command line as:

ngspice -b amprfclsf.cir > out

Load and/or source pull schemes are easily integrated into the SPICE [1–4] transient analysis easily by modifying the load and/or source impedances and by

Table 3.7   Load pull and efficiency for class F amplifier with third harmonic tuning at 500 MHz

Load impedance 7.812 ohm			Source impedance 3.16 ohm			
RMS load current (A)	RMS load voltage (V)	RMS load power (watts)	RMS power supply current (A)	RMS power supply voltage (V)	RMS supplied power (watts)	Efficiency
0.650	5.079	3.30135	0.744	12.5	9.3	0.3549
Load impedance 15.624 ohm			Source impedance 3.16 ohm			
0.539	8.417	4.536763	0.657	12.5	8.2125	0.5524
Load impedance 31.248 ohm			Source impedance 3.16 ohm			
0.441	13.771	6.073911	0.609	12.5	7.6125	0.7977

executing the Ngspice 31 [1] simulator iteratively, for each selected value of the load and/or source impedances. The SPICE transient analysis results are analyzed with the utility C computer language executable *rmscalc*, and the RMS (root mean square) output power and input power are computed, as listed in Table 3.7.

The interested reader is encouraged to try to boost the efficiency even higher, using a combination of the load and/or source pull schemes.

3.9   750 MHz Class F Amplifier with Third, Fifth Harmonic Tuning: Design with SPICE and Efficiency Boost from 27% to 62% with Load Pull

The supplied C computer language [7] executable *amplifierrfclsf* can also be used to generate the text SPICE [1–4] netlist for class F amplifier with both third and fifth harmonic tuning, using the simple command line argument list:

./amplifierrfclsf b ss 750 10 12.5 5

The generated text SPICE [1–4] netlist is listed below:

.INCLUDE hiperftransistormodels

** IMPEDANCE MATCHING AT START, END MUST BE ADDED - LOAD, SOUTCE PULL

** CLASS F RF AMPLIFIER WITH 3RD, 5TH
** HARMONIC TUNING SINGLE SIDED

** RESONATOR C, L TO BE ESTIMATED
** FROM LC PRODUCT VALUE IN PARAMETER LIST

** IMPEDANCE MATCHING TO 50.0 OHM
** MUST BE ADDED SEPARATELY

** RESONATOR C, L FOR 3rd, 5th
** HARMONICS TO BE ESTIMATED
** FROM LC PRODUCT FOR VALUE
** FUNDAMENTAL - 3RD HARMONIC FREQUENCY
** IS 3xFUNDAMENTAL ...

.PARAMS CDCBLK=2.666667e-10 LRFC=2.123142e-08
+ LCPROD=4.507733e-20 LCPROD3=5.008592e-21
+ LCPROD5=7.212373e-21

.PARAMS TS=1.333333e-11 FREQ=7.500000e+08
+ RL=31.249

.PARAMS CR=1.0e-15 LR=4.507733e-5
+ CR3=1.0e-8 LR3=5.008592e-13
+ CR5=1.0e-8 LR5=7.212373e-13
+ AMPL=7.5 LT=1.0e-9 R0=50.0 VCC=12.5

** PULSED INPUT
.PARAMS PER=1.333333e-09 PW=6.666667e-10
+ TR=6.666667e-11 DLY=6.666667e-10

** INITIAL ESTIMATED LOAD IMPEDANCE 7.812 RS=4.71

C0 6 0 1.0e-3
CDCBLK1 7 9 {CDCBLK}
CR 11 0 {CR}
CR3 9 10 {CR3}
CR5 10 11 {CR5}
LRFC 6 7 {LRFC}
LR 11 0 {LR}
LR3 9 10 {LR3}
LR5 10 11 {LR5}
** INPUT TRANSFORMER
LT0 2 0 {LT}
LT1 4 0 {LT}
k0 LT0 LT1 0.99
RL 12 0 {RL}
XQ0 7 8 0 BFR92A
VCC 1 0 DC {VCC} AC 0.0
VSIGP 3 0 DC 0.001 PULSE(0 {AMPL} 0 {TR} {TR} {PW} {PER})
VSIGM 2 3 DC 0.001 PULSE(0 {-AMPL} {DLY} {TR} {TR} {PW} {PER})
VTST0 1 6 DC 0.0 AC 0.0
VTST1 4 8 DC 0.0 AC 0.0
VTST2 11 12 DC 0.0 AC 0.0

.OPTIONS METHOD=GEAR NOPAGE RELTOL=1m
.IC
.TRAN {TS} 70.0us 50.0us UIC

Table 3.8 Load and/or source pull and efficiency for third and fifth harmonic tuned class F amplifier at 750 MHz

Load harmonic 7.812 ohm			Source impedance 4.71 ohm			
RMS load current (A)	RMS load voltage (V)	RMS load power (watts)	RMS power supply current (A)	RMS power supply voltage (V)	RMS supplied power (watts)	Efficiency
0.537	4.192	2.25114	0.662	12.5	8.275	0.2720
Load harmonic 15.624 ohm			Source impedance 4.71 ohm			
0.450	7.038	3.1671	0.591	12.5	7.3875	0.4287
Load harmonic 31.248 ohm			Source impedance 4.71 ohm			
0.349	10.904	3.805496	0.488	12.5	6.1	0.6235

```
** OUTPUT
*.PRINT TRAN V(8) I(VTST1)
** DC SUPPLY CURRENT
.PRINT TRAN I(VTST0)
.END
```

Using identical analysis technique as before, with load and/or source pull integrated into the SPICE [1–4] netlist, followed by estimation of RMS output power and input DC power, the values of efficiency for each combination of load and source impedances are summarized in Table 3.8. The capacitor and inductor values for the three resonators (third and fifth harmonic and fundamental) are estimated using the same reasoning as used in the case of the third harmonic tuned class F amplifier at the operating frequency 500 MHz (previous design example in Sect. 3.8).

The final efficiency value computed for the 750 MHz amplifier is significantly lower than the final efficiency value for the 500 MHz amplifier, with similar load impedances. It is left as an exercise for the interested reader to find out the reason for this.

3.10 500 MHz Class G Amplifier: Design with SPICE and Efficiency Boost from 22% to 81% with Load and Source Pull

The supplied C computer language [7] executable **amplifierrfclsg** generates the text SPICE input format netlist for the class G amplifier, which is a class AB amplifier with special circuitry to switch the positive and negative power supply rails, as the magnitude of the input signal amplitude changes. Both positive and negative polarity power supply rails are needed, as both NPN [5] and PNP [6] bi-junction RF transistors are used. Therefore, the input parameters supplied to *amplifierrfclsg* are identical to those used for the class AB amplifier, and the following command line argument list generates the text SPICE netlist for the 500 MHz class G amplifier.

./amplifierrfclsg C 65 20 14 -14 500 5 12.5 -10

The generated text SPICE [1–4] input format netlist, named *amprfclsg.cir*, is listed below:

.INCLUDE hiperftransistormodels
.INCLUDE diodemodelsgen

** IMPEDANCE MATCHING AT START, END MUST BE ADDED - LOAD, SOURCE PULL

** EDIT|MODIFY PARAMETERS AS NEEDED
** USES BFR92A(NPN) AND BFT92(PNP) RF
** COMPLIMENTARY TRANSISTOR PAIR
** IMPEDANCE MATCHING MUST BE ADDED

.PARAMS VCCNPN=15.250 RCNPN=1089.286
+ RBNPN=70803.571

.PARAMS VCCPNP=-12.750 RCPNP=910.714
+ RBPNP=18214.286
.PARAMS CDCBLK=2.000000e-09 TS=1.000000e-10
.PARAMS FREQ=5.000000e+08 RL=65.023 AMPL=8.0
+ C0=1.0e-6 R=30.0

B0 2 1 V='V(1) >= {(0.5)*VCCNPN} && V(1) > 0.0 ? {VCCNPN} : {(0.5)*VCCNPN}'
B1 3 1 V='V(1) >= {(0.5)*VCCPNP} && V(1) <= 0.0 ? {VCCPNP} : {(0.5)*VCCPNP}'
C0 2 0 {C0}
C1 2 0 {C0}
C2 2 0 {C0}
C3 2 0 {C0}
C4 2 0 {C0}
C5 3 0 {C0}
C6 3 0 {C0}
C7 3 0 {C0}
C8 3 0 {C0}
C9 3 0 {C0}
CDCIN 6 11 {CDCBLK}
CDCOUT 12 13 {CDCBLK}
D0 9 11 1N4007
D1 11 10 1N4007
LRFC0 4 15 1.0e-6
LRFC1 5 16 1.0e-6
RBP 15 9 {RBNPN}
RBM 16 10 {RBPNP}
RCP 15 7 {RCNPN}
RCM 16 8 {RCPNP}
RS 1 6 {R}
RL 14 0 {RL}

```
XQ0 7 9 12 BFR92A
XQ1 8 10 12 BFT92

VSIG 1 0 DC 0.001 SIN(0 {AMPL} {FREQ} 0 0 0)
VTST0 2 4 DC 0.0 AC 0.0
VTST1 3 5 DC 0.0 AC 0.0
VTST2 13 14 DC 0.0 AC 0.0

.OPTIONS METHOD=GEAR NOPAGE RELTOL=1m
.IC
.TRAN {TS} 100.0us 80.0us UIC
** DIAGNOSTIC
* .PRINT TRAN V(?)
** OUTPUT
*.PRINT TRAN V(14) I(VTST2)
** +VE POWER SUPPLY CURRENT - VOLTAGE
*.PRINT TRAN V(4) I(VTST0)
** -VE POWER SUPPLY CURRENT - VOLTAGE
.PRINT TRAN V(5) I(VTST1)
.END
```

The required positive and negative polarity rail switching is achieved with the SPICE arbitrary, user-defined current or voltage source, the 'B' source. Load and/or source pull is easily integrated into the SPICE [1–4] netlist by modifying the load and source impedances and by iteratively executing the Ngspice 31 [1] simulator with each test value. Since both positive and negative polarity power sources are used, the supplied power is the average of the calculated RMS power supplied by the positive and the negative polarity power supplies, respectively. The results of the load and/or source pull experiments are listed in Table 3.9.

Table 3.9 Load and source pull and efficiency for class G amplifier at 500 MHz

Load impedance 16.256			Source impedance 50.0 ohm			
RMS load current (A)	RMS load voltage (V)	RMS load power (watts)	RMS power supply current (A)	RMS power supply voltage (V)	RMS supplied power (watts)	Efficiency
0.03	0.481	0.01443	0.006	12.534	0.07521	
			0.006	9.029	0.054174	0.2230
Load impedance 32.512 ohm			Source impedance 50.0 ohm			
0.03	0.972	0.02916	0.006	12.266	0.073596	
			0.006	9.245	0.05547	0.4518
Load impedance 65.024 ohm			Source impedance 50.0 ohm			
0.027	1.743	0.047061	0.006	12.081	0.072336	
			0.006	9.245	0.05547	0.7363
Load impedance 65.024 ohm			Source impedance 30.0 ohm			
0.028	8.848	0.051744	0.006	12.256	0.072336	
			0.006	9.292	0.055752	0.8079

The reader is encouraged to experiment with the above SPICE [1–4] netlist and boost the efficiency higher.

3.11   500 MHz Class H Amplifier: Design with SPICE and Efficiency Boost from 30% to 84% with Load Pull

The supplied C computer language [7] executable ***amplifierrfclsh*** generates the text SPICE [1–4] netlist for a class H amplifier that is very similar to the class G amplifier. *The difference is that while for the class G amplifier both the positive and negative power supply rails are switched between discrete voltage levels (as the input signal magnitude changes with time),* **the class H circuitry modulates the power supply rails as the input signal magnitude changes.** Both class G and H amplifiers use the class AB amplifier, only the scheme to customize and/or optimize the power supply rails, to enhance efficiency, is different. The simple command line argument list

./amplifierrfclsh b 65 20 14 -14 500 5 12.5 -10

generates the text SPICE [1–4] input format netlist amprfclsh.cir, which is almost identical to the SPICE [1–4] netlist for the class G amplifier, except the way the positive and negative power supply rails are adjusted with reference to the input signal.

B0 2 1 V='V(1) >= {(0.5)*VCCNPN} && V(1) > 0.0 ? {1.5*V(1)} : {(0.5)*VCCNPN}'
B1 3 1 V='V(1) >= {(0.5)*VCCPNP} && V(1) <= 0.0 ?{(1.5)*V(1) : {(0.5)*VCCPNP}'

Just like the class G amplifier, the load and/or source pull techniques are easily integrated into the SPICE [1–4] netlist, and by iteratively executing the SPICE simulator for each combination of the load and source impedance, the optimum efficiency can be determined, as listed in Table 3.10.

The reader is encouraged to design and evaluate with SPICE [1–4] a simple circuit to perform the power supply rail modulation for a class H amplifier.

3.12   750 MHz Class E Amplifier: Design with SPICE and Modified Load Source Pull to Boost Efficiency

The C computer language [7] executable ***amplifierrfclse*** generates the text SPICE [1–4] input format netlist, given the designer-specified amplifier parameters as operating frequency, target load output power, and loaded quality factor (Q_L and bi-junction transistor collector-emitter voltage. Sokal's original treatment of the

Table 3.10 Load and source pull impedance and efficiency for class H amplifier at 500 MHz

Load impedance 16.256 ohm			Source impedance 50.0 ohm			
RMS load current (A)	RMS load voltage (V)	RMS load power (watts)	RMS power supply current (A)	RMS power supply voltage (V)	RMS supplied power (watts)	Efficiency
0.029	0.469	0.013601	0.005	11.010	0.05505	
			0.005	7.331	0.036605	0.2968
Load impedance 32.512 ohm			Source impedance 50.0 ohm			
0.027	0.842	0.022734	0.005	11.488	0.05744	
			0.005	7.249	0.036245	0.4853
Load impedance 65.024 ohm			Source impedance 50.0 ohm			
0.025	1.573	0.039325	0.005	11.022	0.05511	
			0.005	7.681	0.038405	0.8409

class E amplifier (Chap. 2, Sect. 2.3.7) was based entirely on numerical analysis (specifically nonlinear curve-fit), and all class E amplifier circuit passive component values were expressed as functions of the circuit loaded quality factor Q_L. *Thus, the original load and source pull schemes, applied successfully so far to boost the efficiencies of the other classes of amplifiers, cannot be directly applied in the case of the class E amplifier.* The only way to boost efficiency is to ensure that the shunt capacitor across the transistor collector/drain and emitter/source is not simultaneously getting charged and discharged. This is tricky, because the capacitance value varies with Q_L and larger value capacitors take longer to charge and discharge.

The scheme to analyze the class E amplifier is to vary the loaded quality factor Q_L at a fixed operating frequency and measure the efficiency for each value of Q_L. amplifierrfclsesokal is invoked easily from the command line, and the RMS output power and the RMS power from the power source are measured each time, with the supplied C computer language [7] utility **rmscalc**. As per Sokal's original design and analysis of the class E amplifier, the minimum possible value for Q_L is 1.79. The command line argument to create the text SPICE [1–4] input format netlist for a class E amplifier with $Q_L = 2$, operating at 750 MHz with target load power of 10 watts with NPN RF transistor BFR92A, is:

./amplifierrfclsesokal b SS x 750 10 2 12.5

The generated text SPICE [1–4] input format netlist is as follows:

.INCLUDE hiperftransistormodels

** IMPEDANCE MATCHING AT START, END MUST BE ADDED - LOAD, SOURCE PULL

** ADJUST PARAMETERS AS NECESSARY TO
** OPTIMIZE AMPLIFIER PERFORMANCE

```
.PARAMS CAPSHUNT=7.951448e-12 CR=1.065771e-10
+ LR=2.580041e-09 RL=6.076
.PARAMS TS=6.666667e-11 FREQ=7.500000e+08
.PARAMS VCC=12.500 AMPL=7.5 LRFC=1.000000e-07
.PARAMS PER=1.333333e-09 PW=6.666667e-10
+ TR=8.888889e-11


** Loaded Q 2.000


C0 3 0 1.0e-5
CAPSHUNT 5 0 {CAPSHUNT}
CR 6 7 {CR}
LR 5 6 {LR}
** INPUT TRANSFORMER
LT0 2 0 1.0e-9
LT1 4 0 1.0e-9
k0 LT0 LT1 0.99
LRFC1 3 5 {LRFC}
XQ0 5 11 0 BFR92A
RL 8 0 {RL}
VCC 1 0 DC {VCC} AC 0.0
VSIG 2 0 DC 0.001
 SIN(0 {AMPL} {FREQ} 0 0 0)


** PULSE TRIBBER
** VSIG 2 0 DC 0.001 PULSE(0 {0.85*VCC} 0 {TR} {TR} {PW} {PER})
VTST0 1 3 DC 0.0 AC 0.0
VTST1 4 11 DC 0.0 AC 0.0
VTST2 7 8 DC 0.0 AC 0.0


.OPTIONS METHOD=GEAR NOPAGE RELTOL=1m
.IC
.TRAN {TS} 70.0us 50.0us UIC
** OUTPUT
.PRINT TRAN V(8) I(VTST2)
** INPUT
*.PRINT TRAN V(11) I(VTST1)
**COLLECTOR CURRENT, CoLLECTOR-EMITTER VOLTAGE
*.PRINT TRAN I(VTST0) ;V(5)
.END
```

The text SPICE [1–4] netlists for the same amplifier with different Q_L are created with similar command line argument lists. In each case, the SPICE [1–4] transient analysis results are processed with *rmscalc.* .The results are tabulated below (Table 3.11).

This design example demonstrates the peculiar behavior of the class E RF power amplifier. At high operating frequencies, the efficiency is low at start, and as the load pull scheme is applied, the efficiency falls. The reader is encouraged to determine the reason for this.

Table 3.11 Quality factor and efficiency at 750 MHz

RMS load current (A)	RMS load voltage (V)	RMS load power (watts)	RMS power supply current (A)	RMS power supply voltage (V)	RMS power delivered (watts)	Efficiency
$Q_L = 2.0$load impedance 6.076 ohm			Source impedance 4.71 ohm			
0.568	3.45	1.9596	0.399	12.5	4.9875	0.3929
$Q_L = 4.0$load impedance 7.783 ohm			Source impedance 4.71 ohm			
0.466	3.627	1.690182	0.433	12.5	5.4125	0.3123
$Q_L = 6.0$load impedance 8.254 ohm			Source impedance 4.71 ohm			
0.444	3.669	1.628592	0.4	12.5	5.0	0.3257
$Q_L = 8.0$load impedance 8.68 ohm			Source impedance 4.71 ohm			
0.437	3.703	1.618211	0.412	12.5	5.15	0.3142
$Q_L = 10.0$load impedance8.865 ohm			Source impedance 4.71 ohm			
0.435	3.736	1.62516	0.404	12.5	5.05	0.3218

3.13   500 MHz Doherty Amplifier Quadrature Coupler/Splitter and Amplifier Output Combiner

The Doherty amplifier is a cleaver combination of three sub-circuits, the quadrature coupler/splitter, the middle parallel combination of the carrier (class B), and peaking (class C) amplifiers followed by the signal combiner. As both the class B and C design examples have been examined and analyzed in minute detail in previous sections, only the quadrature coupler and signal combiner are examined in detail here.

The following hand-crafted SPICE [1–4] netlist enables the Ngspice simulator to perform the transient analysis of the *quadrature coupler/splitter and the signal combiner, with a high peak to average ratio input signal source.*

QUADRATURE SPLITTER AND COMBINER

.PARAMS FREQ=5.000000e+08 WLEN=6.000000e-01 LEN=1.500000e-01 TD=5.000000e-10

.PARAMS R=50.000 TS=1.000000e-10 AMPL=7.5
.PARAMS PER=2.000000e-08 PW=2.000000e-09 TR=3.000000e-11

** IDEAL TX. LINE

.SUBCKT QSPLTI 1 2 3 4
** 1 IN 1

```
** 2 IN 2
** 3 OUT 1
** 4 OUT 2
T0 1 0 3 0 Z0={0.707*R} TD={TD}
T1 1 0 2 0 Z0={R} TD={TD}
T2 2 0 4 0 Z0={0.707*R} TD={TD}
T3 3 0 4 0 Z0={R} TD={TD}
.ENDS

.SUBCKT QCMBI 1 2 3
** 1 IN 1
** 2 IN 2
** 3 OUT
T0 1 0 3 0 Z0={0.707*R} TD={TD}
T1 2 0 1 0 Z0={R} TD={TD}
.ENDS

RS 1 2 {R}
RG 3 0 {R}
RSP0 4 0 {R}
RSP1 5 0 {R}
RL 6 0 {R}
XQS 2 3 4 5 QSPLTI
XCMB 4 5 6 QCMBI
VSIG0 7 0 DC 0.001
+ SIN(0 {AMPL} {FREQ} 0 0 0)
VSIG1 1 7 DC 0.001 PULSE(0 {5.0*AMPL} 0 {TR} {TR} {PW} {PER})

.OPTIONS METHOD=GEAR NOPAGE RELTOL=1m
.IC
.TRAN {TS} 35.0us 30.0us UIC
** DIRECT, QUADRATURE SPLITS
*.PRINT TRAN V(4) V(5)
** OUTPUT
.PRINT TRAN V(6)
.END
```

As in the case of all previous design examples, the Ngspice 31 simulator performs transient analysis with this netlist for two sets of outputs – the quadrature coupler/splitter and the combiner. These outputs are shown in Fig. 3.1a, b.

It is left to the reader to insert the appropriate class B and C netlist into the above netlist and do the SPICE [1–4] transient analysis for the full Doherty amplifier. After that load and/or source pull analyses can be performed.

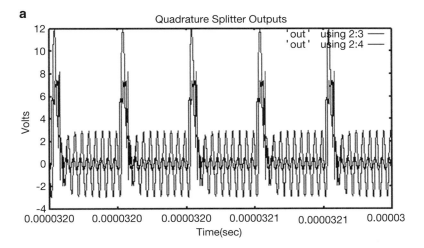

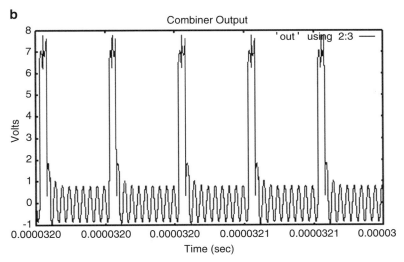

Fig. 3.1 (**a**) Quadrature coupler/splitter output. (**b**) Signal combiner output. The quadrature coupler/splitter output is input to the signal combiner

3.14 500 MHz Class A Amplifier: SPICE Input, Output Noise Spectra Measurement

The built-in electronic noise spectrum measurement feature of SPICE [1–4] can be used easily, simply by editing/modifying the corresponding SPICE [1–4] netlist for any RF power amplifier. So the SPICE [1–4] netlist for the class A RF power amplifier (Sect. 3.2) is modified as follows:

.INCLUDE hiperftransistormodels

** CLASS A AMPLIFIER WITH BASE BIAS

** INPUT|OUTPUT IMPEDANCE MATCHING MUST BE ADDED - LOAD SOURCE
PULL

.PARAMS VCC=25.000 RC=892.857 RE=178.571 RB0=11607.143 RB1=1703.801
.PARAMS CDCBLK=2.000000e-09 CEBYP=1.783439e-13 TS=1.000000e-10 RL=62.500
R=50.0
.PARAMS FREQ=5.000000e+08 AMPL=18.750 C0=1.0e-3 LRFC=1.0E-6

** NOISE ANALYSIS

.PARAMS FLLIM=5.000000e+07 FHLIM=7.500000e+08 TOT=25000

** INITIAL ESTIMATE RL=62.500 R=50.0

C0 3 0 {C0}
CDCBLK0 7 4 {CDCBLK}
CDCBLK1 5 8 {CDCBLK}
CEBYP 6 0 {CEBYP}
LRFC 3 11 {LRFC}
RB1 11 4 {RB0}
RB2 4 0 {RB1}
RC 11 5 {RC}
RE 6 0 {RE}
RS 2 7 {R}
RL 10 0 {RL}
XQ0 5 9 6 BFR92A

VCC 1 0 DC {VCC} AC 0.0
** TRANSIENT ANALYSIS
*VSIG 2 0 DC 0.002
*+ SIN(0 {AMPL} {FREQ} 0 0 0)

** NOISE ANALTSIS
VSIG 2 0 DC 0.002
 AC 1.0 SIN(0 {AMPL} {FREQ} 0 0 0)

VTST0 1 3 DC 0.0 AC 0.0
VTST1 4 9 DC 0.0 AC 0.0
VTST2 8 10 DC 0.0 AC 0.0

.OPTIONS METHOD=GEAR NOPAGE RELTOL=1m
** NOISE ANALYSIS
.NOISE V(10) VSIG LIN {TOT} {FLLIM} {FHLIM}

*.IC
*.TRAN {TS} 50.0us 30.0us UIC
** OUTPUT

```
*.PRINT TRAN V(10) I(VTST2)
** INPUT
*.PRINT TRAN V(9) I(VTST1)
** DC SUPPLY CURRENT
*.PRINT TRAN I(VTST0)

** INPUT, OUTPUT NOISE SPECTRUM
.CONTROL
LISTING L
RUN
WRITE amprfclsaall.raw noise1.all noise2.all
PRINT V(inoise_total) V(onoise_total)
QUIT
.ENDC

.END
```

In the above SPICE [1–4] netlist, the sections corresponding to transient analysis have been commented out to perform the input and output noise spectrum analyses. When executed in the batch/command line argument mode, the Ngspice 31 simulator generates the output:

```
ngspice -b amprfclsa.cir
Circuit: *INCLUDE hiperftransistormodels
 *INCLUDE hiperftransistormodels
 2 : .param vcc=25.000
 3 : .param rc=892.857
 4 : .param re=178.571
 5 : .param rb0=11607.143
 6 : .param rb1=1703.801
 7 : .param cdcblk=2.000000e-09
 8 : .param cebyp=1.783439e-13
 9 : .param ts=1.000000e-10
 10 : .param rl=62.500
 11 : .param r=50.0
 12 : .param freq=5.000000e+08
 13 : .param ampl=18.750
 14 : .param c0=1.0e-3
 15 : .param lrfc=1.0e-6
 16 : .param fllim=5.000000e+07
 17 : .param fhlim=7.500000e+08
 18 : .param tot=25000
 19 : .global gnd
. . . .....
. . . .....
. . . .....
 170 : c0 3 0 {c0}
 171 : cdcblk0 7 4 {cdcblk}
 172 : cdcblk1 5 8 {cdcblk}
 173 : cebyp 6 0 {cebyp}
 174 : lrfc 3 11 {lrfc}
```

```
175 : rb1 11 4 {rb0}
176 : rb2 4 0 {rb1}
177 : rc 11 5 {rc}
178 : re 6 0 {re}
179 : rs 2 7 {r}
180 : rl 10 0 {rl}
181 : xq0 5 9 6 bfr92a
182 : vcc 1 0 dc {vcc} ac 0.0
187 : vsig 2 0 dc 0.002 ac 1.0 sin(0 {ampl} {freq} 0 0 0)
188 : vtst0 1 3 dc 0.0 ac 0.0
189 : vtst1 4 9 dc 0.0 ac 0.0
190 : vtst2 8 10 dc 0.0 ac 0.0
193 : .noise v(10) vsig lin {tot} {fllim} {fhlim}
203 : .control
204 : listing l
205 : run
206 : write amprfclsaall.raw noise1.all noise2.all
207 : print v(inoise_total) v(onoise_total)
208 : quit
209 : .endc
191 : .options method=gear nopage reltol=1m
211 : .end
```

Doing analysis at TEMP = 27.000000 and TNOM = 27.000000
Reference value : 4.42632e+08

No. of Data Rows : 25000
No. of Data Rows : 1
v(inoise_total) = 1.090476e-04
v(onoise_total) = 3.524005e-05
ngspice-31 done

Note that the load and source impedances have not been optimized with load and source pull. The input and output noise spectra are written to a binary file **amprfclsaall.raw,** which is in XRD format (X ray diffraction), and therefore cannot be opened or its contents displayed with the popular desktop graphics package Gnuplot. This has to be performed interactively using SPICE, as shown below:

```
$ ngspice
******
** ngspice-31 : Circuit level simulation program
** The U. C. Berkeley CAD Group
** Copyright 1985-1994, Regents of the University of California.
** Please get your ngspice manual from
http://ngspice.sourceforge.net/docs.html
** Please file your bug-reports at
http://ngspice.sourceforge.net/bugrep.html
******
ngspice 1 -> load amprfclsaall.raw
Loading raw data file ("amprfclsaall.raw") . . . done.
Title: *INCLUDE hiperftransistormodels
Name: Noise Spectral Density Curves
```

Date: Fri Apr 24 10:23:31 2020
Title: *INCLUDE hiperftransistormodels
Name: Integrated Noise
Date: Fri Apr 24 10:23:32 2020
Here are the vectors currently active:
Title: *INCLUDE hiperftransistormodels
Name: noise2 (Integrated Noise)
Date: Fri Apr 24 10:23:32 2020
 inoise_total : voltage, real, 1 long
 onoise_total : voltage, real,
-- hit return for more, ? for help --
 1 long [default scale]
ngspice 2 -> set color0 = rgb:f/f/e
ngspice 3 -> set color1 = rgb:1/1/1
ngspice 4 -> setplot noise1
ngspice 5 -> display
Here are the vectors currently active:
Title: *INCLUDE hiperftransistormodels
Name: noise1 (Noise Spectral Density Curves)
Date: Fri Apr 24 10:23:31 2020
 frequency : frequency, real, 25000 long [default scale]
-- hit return for more, ? for help --
 inoise_spectrum : notype, real, 25000 long
 onoise_spectrum : notype, real, 25000 long
ngspice 6 -> plot inoise_spectrum
ngspice 7 ->
[3~: no such command available in ngspice
ngspice 8 -> plot onoise_spectrum
ngspice 9 -> exit
ngspice-31 done

The generated input and output noise spectra are shown in Fig. 3.2a, b.

3.15   300 MHz Distributed Amplifier Design with SPICE

The C computer language [7] executable **amplifierrfdist** generates the SPICE [1–4] netlist for a 30-stage distributed amplifier (CA_ using the 2N4416 N-type junction field effect transistor (N JFET)). The designer supplies the frequency band center frequency (f_C) and the characteristic impedance (Z_0) of the transmission line segments (50.0 ohm for RF and microwave applications). Another JFET may be used, but the user has to check the data sheet for the selected device. **amplifierrfdist** is invoked from the command line, and to get help information, simply type **./ amplifierrfdist** at the Linux shell command prompt. For this particular design, the command line argument list is as follows:

./amplifierrfdist b 300 50

a

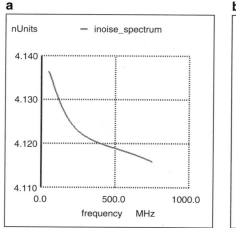

b

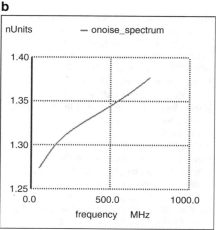

Fig. 3.2 (**a**) 500 MHz class A amplifier input noise spectrum. (**b**) 500 MHz class A amplifier output noise spectrum

The generated SPICE [1–4] netlist named **amprfdist.cir** is listed here:

```
.INCLUDE jfetmodels

** USES N JFET 2N4416

.PARAMS FREQ=3.000000e+08 CAPDCBLK=3.333333e-09
+ WVLEN=1.000000e+00 LEN=2.500000e-01 INDVAL=6.634820e-09
.PARAMS TD=8.333333e-10 TS=1.666667e-10 LRFC=5.307856e-08
+ Z0=50.000
.PARAMS FLLIM=3.000000e+07 FHLIM=4.500000e+08 TOT=40000
.PARAMS AMPL=8.0 VDD=14.0

.SUBCKT DAS 1 2 3 4
** 1 IN 1
** 2 IN 2
** 3 OUT 1
** 4 OUT 1
JFN 1 2 0 2N4416
T0 1 0 3 0 Z0={Z0} TD={TD}
T1 2 0 4 0 Z0={Z0} TD={TD}
.ENDS
.SUBCKT DASL 1 2 3 4
** 1 IN 1
** 2 IN 2
** 3 OUT 1
** 4 OUT 1
JFN 1 2 0 2N4416
```

```
L0 1 3 {INDVAL}
L1 2 4 {INDVAL}
.ENDS

.SUBCKT DASAMP 1 2 3 4
** 1 IN 1
** 2 IN 2
** OUT 1
** OUT 2
XDAS0 1 2 5 6 DAS
XDAS1 5 6 7 8 DAS
XDAS2 7 8 9 10 DAS
XDAS3 9 10 11 12 DAS
XDAS4 11 12 13 14 DAS
XDAS5 13 14 15 16 DAS
XDAS6 15 16 17 18 DAS
XDAS7 17 18 19 20 DAS
XDAS8 19 20 21 22 DAS
XDAS9 21 22 23 24 DAS
XDAS10 23 24 25 26 DAS
XDAS11 25 26 27 28 DAS
XDAS12 27 28 29 30 DAS
XDAS13 29 30 31 32 DAS
XDAS14 31 32 33 34 DAS
XDAS15 33 34 35 36 DAS
XDAS16 35 36 37 38 DAS
XDAS17 37 38 39 40 DAS
XDAS18 39 40 41 42 DAS
XDAS19 41 42 43 44 DAS
XDAS20 43 44 45 46 DAS
XDAS21 45 46 47 48 DAS
XDAS22 47 48 49 50 DAS
XDAS23 49 50 51 52 DAS
XDAS24 51 52 53 54 DAS
XDAS25 53 54 55 56 DAS
XDAS26 55 56 57 58 DAS
XDAS27 57 58 59 60 DAS
XDAS28 59 60 61 62 DAS
XDAS29 61 62 3 4 DAS
.ENDS

CAPADD 4 5 {CAPDCBLK}
CAPDCBLK1 2 6 {CAPDCBLK}
CAPDCBLK2 8 10 {CAPDCBLK}
LRFC 3 4 {LRFC}
RL 11 0 {Z0}
RT0 5 0 {Z0}
RT1 9 0 {Z0}
XDASAMP0 4 7 8 9 DASAMP

VDD 1 0 DC {VDD} AC 0.0
** TRANSIENT ANALYSIS
VSIG 2 0 DC 0.001 SIN(0 {AMPL} {FREQ} 0 0 0)
```

```
** AC(SMALL SIGNAL) ANALYSIS
*VSIG 2 0 DC 0.001 AC {0.1*AMPL}
VTST0 1 3 DC 0.0 AC 0.0
VTST1 6 7 DC 0.0 AC 0.0
VTST2 10 11 DC 0.0 AC 0.0


.OPTIONS METHOD=GEAR NOPAGE RELTOL=1m
** TRANSIENT ANALYSIS
.IC
.TRAN {TS} 400.0ns 25.0ns UIC
** OUTPUT VOLTAGE CIRRENT
.PRINT TRAN V(11) I(VTST2)
** INPUT VOLTAGE CURRENT
*.PRINT TRAN V(7) I(VTST1)
** POWER SUPPLY CURRENT
*.PRINT TRAN I(VTST0)


** AC(SMALL SIGNAL) ANALYSIS
*.AC LIN {TOT} {FLLIM} {FHLIM}
*.PRINT AC V(11)


.END
```

As in the previous design examples, appropriate sections of the SPICE [1–4] may be commented out to analyze transient (steady state) or AC (small signal) performance characteristics of the distributed amplifier. *It is worth noting that by design, impedance matching at the input and output ports has been accounted for, and there is no need for any load/source pull design space exploration.* The transient analysis voltage across the load impedance and the corresponding current through it is shown in Fig. 3.3a, b, respectively.

The supplied utility C computer language [7] executable rmscalc is used to compute the RMS (root mean square) current and voltage at the input and output ports. The RMS output voltage and RMS output current are **9.252 0.185 207090**, where the total number of data samples is 207,090. Similarly, the RMS input voltage and current are **6.084 0.138 207090** and the RMS power supply current is **0.514 A**. These are summarized in Table 3.12.

Therefore, the gain is 2.03863305 or 3.093390606 dB. The efficiency is 25%, which is mediocre compared to the RF power amplifiers that have been examined in detail before. The reason behind the low efficiency is that the single power supply is supplying electrical power to all 30 FETs.

Exercises

• For each of the RF amplifiers analyzed in this chapter, design appropriate input| output L or transmission line impedance matching transformers to match the optimized input|output impedances with the desired ones. For example, for the class A amplifier (Sect. 3.2), the output impedance corresponding to 40% efficiency has been determined to be 500.0 ohm. But the desired output impedance is 50.0 ohm. So an appropriate L or transmission line impedance matching

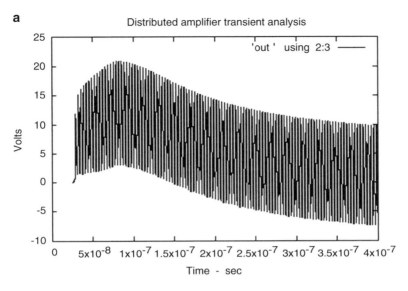

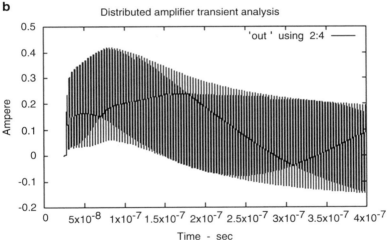

Fig. 3.3 (**a**) Distributed amplifier output voltage across load impedance. (**b**) Distributed amplifier output current through load impedance

transformer must be inserted such that the output impedance, as it appears to the amplifier output port, is 50.0 ohm. Same arguments hold for the input port.

- Use *amplifierrfclsc* to generate the SPICE netlist for a class C fixed-base bias amplifier. Analyze this generated netlist with any available SPICE simulator, and use the load and source pull schemes to boost the efficiency from its starting value,

Table 3.12 Distributed amplifier RMS load, input current, voltage, and RMS power

RMS load voltage (V)	RMS load current (A)	RMS load power (watts)
9.252	0.185	1.71162
RMS input voltage (V)	RMS input current (A)	RMS input power (watts)
6.084	0.138	0.839592
Power supply voltage (V)	RMS power supply Current (A)	RMS supplied power (watts)
14.0	0.514	7.196

- Using *amplifierrfclsd*, design a class D single-sided amplifier at an RF frequency of your choice and analyze its efficiency with load and/or source pull.
- Explain why, despite using identical load impedance values, the final efficiency value computed with the transformer controlled current source (TCCS) class D amplifier is lower than the corresponding final efficiency value computed with the TCVS design example.
- Explain why achieving high efficiency with the class E amplifier is very difficult in practice.
- Explain why both design examples for the class F amplifier use pulsed signal input.
- Design and run the SPICE simulation of a simple electronic circuit to perform power supply rail switching for a class G amplifier.
- Design single-sided class B and C amplifiers to operate at 500 MHz, and insert these in between the quadrature coupler/splitter and signal combiner for the Doherty amplifier. Then use SPICE transient analysis to evaluate the output of the full Doherty amplifier.
- Estimate the input and output noise spectra for the class A RF power amplifier in Sect. 3.14 does not use load- and source-pull-optimized load and source imped-ances. Use the optimized load and source impedance values determined in Sect. 3.2 to perform another SPICE input and output noise analyses of the 500 MHZ class A RF power amplifier of Sect. 3.14.

References

1. Latest Ngspice version 31 user guide and manual from: http://ngspice.sourceforge.net/docs/ngspice-manual.pdf
2. LTSpice users guide and manual from: https://ecee.colorado.edu/~mathys/ecen1400/pdf/scad3.pdf
3. Pspice users guide and manual from: https://www.seas.upenn.edu/~jan/spice/PSpice_UserguideOrCAD.pdf
4. HSpice users guide and manual from: https://cseweb.ucsd.edu/classes/wi10/cse241a/assign/hspice_sa.pdf

5. RF bi-junction NPN transistor BFR92A SPICE model from: http://www.bdtic.com/download/
NXP/spice_BFR92A.prm
6. RF bi-junction PNP transistor BFT92 SPICE model from: https://www.nxp.com/downloads/en/
spice-model/BFT92_SPICE.prm
7. Ebook of the all-time classic C programming language book by the creators of the C computer
language Brian Kernighan and Dennis Ritchie, can be downloaded easily from: http://www2.cs.
uregina.ca/~hilder/cs833/Other%20Reference%20Materials/The%20C%20Programming%
20Language.pdf

Appendices

Appendix A

List of Supplied Software Executables

Two sets of ready-to-use C computer language executables, one each for the Linux and Windows operating systems have been supplied. The executables for the Linux operating system have been compiled on Ubuntu 16.04 LTS, and Fedora 18. All Linux operating system executables at present are 32-bit, and so can easily be executed on any computer using any 64 bit Linux version. Executables exclusively for 64 bit Linux will be available in the near future. The executables for the Windiows operating system come with the ".exe": extension, and work under the very reliable and widely used Linux emulator for Windows – MinGW(*Min*imalist *G*NU For *W*indows). Instructions for downloading and installing the MingDW package on any Windows operating system computer are detailed in Appendix B.

Executable name	Function
amplifierrfclsa	Accepts command line arguments to generate text SPICE input format netlist for any available SPICE simulator – for single sided class A RF power amplifier only
amplifierrfclsb	Accepts command line arguments to generate text SPICE input format netlist for any available SPICE simulator – for single and double sided class B RF power amplifier
amplifierrfclsab	Accepts command line arguments to generate text SPICE input format netlist for any available SPICE simulator – only for double sided class AB RF power amplifier
amplifierrfclsc	Accepts command line arguments to generate text SPICE input format netlist for any available SPICE simulator – only for single sided, fixed and zero base bias class C RF power amplifier

(continued)

Executable name	Function
amplifierrfclsd	Accepts command line arguments to generate text SPICE input format netlist for any available SPICE simulator – only for double sided(TCCS, TCVS) class D RF power amplifier
amplifierrfclse	Accepts command line arguments to generate text SPICE input format netlist for any available SPICE simulator – for single sided class E RF power amplifier only
amplifierrfclsf	Accepts command line arguments to generate text SPICE input format netlist for any available SPICE simulator – for single sided class F RF power amplifier only
amplifierrfclsg	Accepts command line arguments to generate text SPICE input format netlist for any available SPICE simulator – only for double sided class G(really a class AB) RF power amplifier
amplifierrfclsh	Accepts command line arguments to generate text SPICE input format netlist for any available SPICE simulator – only for double sided class H(really a class AB) RF power amplifier
amplifierrfdist	Accepts the band center frequency and characteristic impedance of a candidate transmission line to generate the SPICE nellist for a 30 stage distributed amplifier using the N JFET 2N4416/
rmscalc	Computes RMS(Root Mean Squared) values for a tab separated variable, maximum five(5) columns of data in text format. Number of rows is unlimited

Appendix B

How to Install MinGW(Minimalist GNU for Windows) on Windows Operating System Computers

MinGW is the most reliable, trusted and widely used Linux emulator for Windows (including Windows 10), with the built-in *gcc* compiler suite. It can be easily downloaded and used on any computer running the Windows(7, 8, 10) operating system, The following URLs contain detailed, step-by-step instructions(often accompanied by screenshots) for qonloading and installing MinGW on any Windows operating system computer. In addition, there are a set of YouTube videos that provide the same step-by-step instructions.

http://www.codebind.com/cprogramming/install-mingw-windows-10-gcc/

https://www.rose-hulman.edu/class/csse/resources/MinGW/installation.htm

https://medium.com/@paadddddddyyyyy/how-to-install-mingw-on-windows-10-gcc-8-2-0-9a9074383af9

https://www.instructables.com/id/How-to-Install-MinGW-GCCG-Compiler-in-Windows-XP78/

https://genome.sph.umich.edu/wiki/Installing_MinGW_%26_MSYS_on_WindowsA

https://cpp.tutorials24x7.com/blog/how-to-install-mingw-on-windows

https://www.ics.uci.edu/~pattis/common/handouts/mingweclipse/mingw.html
https://azrael.digipen.edu/~mmead/www/public/mingw/
http://www.mingw.org/wiki/HOWTO_Install_the_MinGW_GCC_Compiler_
Suite
https://www3.ntu.edu.sg/home/ehchua/programming/howto/Cygwin_HowTo.
html

Index

© The Author(s), under exclusive license to Springer Nature Switzerland AG 2021   71
A. Banerjee, *Practical RF Amplifier Design and Performance Optimization*
with SPICE and Load- and Source-pull Techniques,
https://doi.org/10.1007/978-3-030-62512-2

Printed in the United States
by Baker & Taylor Publisher Services